Climate
Systems

GLOBAL WARMING

Climate Systems

Interactive Forces
of Global Warming

Julie Kerr Casper, Ph.D.

☑ Facts On File
An imprint of Infobase Publishing

CLIMATE SYSTEMS: Interactive Forces of Global Warming

Facts On File, Inc.
An imprint of Infobase Publishing
132 West 31st Street
New York NY 10001

Library of Congress-in-Publication Data
Casper, Julie Kerr.
  Climate systems : interactive forces of global warming / Julie Kerr Casper.
     p. cm.—(Global warming)
  Includes bibliographical references and index.
  ISBN 978-0-8160-7260-6
  1. Climatology—Popular works. 2. Global warming—Popular works. I. Title.
QC863.4.C37 2009
551.5—dc22   2008040921

Facts On File books are available at special discounts when purchased in bulk quantities for businesses, associations, institutions, or sales promotions. Please call our Special Sales Department in New York at (212) 967-8800 or (800) 322-8755.

You can find Facts On File on the World Wide Web at http://www.factsonfile.com

Text design by Erik Lindstrom
Illustrations by Jeremy Eagle and Melissa Ericksen
Photo research by the author

Printed in the United States of America

Bang FOF 10 9 8 7 6 5 4 3 2 1

This book is printed on acid-free paper.

CONTENTS

PREFACE

We do not inherit the Earth from our ancestors—
we borrow it from our children.

This ancient Native American proverb and what it implies resonates today as it has become increasingly obvious that people's actions and interactions with the environment affect not only living conditions now, but also those of many generations to follow. Humans must address the effect they have on the Earth's climate and how their choices today will have an impact on future generations.

Many years ago, Mark Twain joked that "Everyone talks about the weather, but no one does anything about it." That is not true anymore. Humans are changing the world's climate and with it the local, regional, and global weather. Scientists tell us that "climate is what we expect, and weather is what we get." Climate change occurs when that average weather shifts over the long term in a specific location, a region, or the entire planet.

Global warming and climate change are urgent topics. They are discussed on the news, in conversations, and are even the subjects of horror movies. How much is fact? What does global warming mean to individuals? What should it mean?

The readers of this multivolume set—most of whom are today's middle and high school students—will be tomorrow's leaders and scientists. Global warming and its threats are real. As scientists unlock the mysteries of the past and analyze today's activities, they warn that future generations may be in jeopardy. There is now overwhelming evidence that human activities are changing the world's climate. For thousands of years, the Earth's atmosphere has changed very little; but today, there are problems in keeping the balance. Greenhouse gases are being added to the atmosphere at an alarming rate. Since the Industrial Revolution (late 18th, early 19th centuries), human activities from transportation, agriculture, fossil fuels, waste disposal and treatment, deforestation, power stations, land use, biomass burning, and industrial processes, among other things, have added to the concentrations of greenhouse gases.

These activities are changing the atmosphere more rapidly than humans have ever experienced before. Some people think that warming the Earth's atmosphere by a few degrees is harmless and could have no effect on them; but global warming is more than just a warming—or cooling—trend. Global warming could have far-reaching and unpredictable environmental, social, and economic consequences. The following demonstrates what a few degrees' change in the temperature can do.

The Earth experienced an ice age 13,000 years ago. Global temperatures then warmed up 8.3°F (5°C) and melted the vast ice sheets that covered much of the North American continent. Scientists today predict that average temperatures could rise 11.7°F (7°C) during this century alone. What will happen to the remaining glaciers and ice caps?

If the temperatures rise as leading scientists have predicted, less freshwater will be available—and already one-third of the world's population (about 2 billion people) suffer from a shortage of water. Lack of water will keep farmers from growing food. It will also permanently destroy sensitive fish and wildlife habitat. As the ocean levels rise, coastal lands and islands will be flooded and destroyed. Heat waves could kill tens of thousands of people. With warmer temperatures, outbreaks of diseases will spread and intensify. Plant pollen mold spores in the air will increase, affecting those with allergies. An increase in severe weather could result in hurricanes similar or even stronger than Katrina in 2005, which destroyed large areas of the southeastern United States.

Higher temperatures will cause other areas to dry out and become tinder for larger and more devastating wildfires that threaten forests, wildlife, and homes. If drought destroys the rain forests, the Earth's delicate oxygen and carbon balances will be harmed, affecting the water, air, vegetation, and all life.

Although the United States has been one of the largest contributors to global warming, it ranks far below countries and regions—such as Canada, Australia, and western Europe—in taking steps to fix the damage that has been done. Global Warming is a multivolume set that explores the concept that each person is a member of a global family who shares responsibility for fixing this problem. In fact, the only way to fix it is to work together toward a common goal. This seven-volume set covers all of the important climatic issues that need to be addressed in order to understand the problem, allowing the reader to build a solid foundation of knowledge and to use the information to help solve the critical issues in effective ways. The set includes the following volumes:

Climate Systems
Global Warming Trends
Global Warming Cycles
Changing Ecosystems
Greenhouse Gases
Fossil Fuels and Pollution
Climate Management

These volumes explore a multitude of topics—how climates change, learning from past ice ages, natural factors that trigger global warming on Earth, whether the Earth can expect another ice age in the future, how the Earth's climate is changing now, emergency preparedness in severe weather, projections for the future, and why climate affects everything people do from growing food, to heating homes, to using the Earth's natural resources, to new scientific discoveries. They look at the impact that rising sea levels will have on islands and other areas worldwide, how individual ecosystems will be affected, what humans will lose if rain forests are destroyed, how industrialization and pollution puts peoples' lives at risk, and the benefits of developing environmentally friendly energy resources.

The set also examines the exciting technology of computer modeling and how it has unlocked mysteries about past climate change and global warming and how it can predict the local, regional, and global climates of the future—the very things leaders of tomorrow need to know *today*.

> *We will know only what we are taught;*
> *We will be taught only what others deem is important to know;*
> *And we will learn to value that which is important.*
> —Native American proverb

ACKNOWLEDGMENTS

Global warming may very well be one of the most important issues you will have to make a decision on in your lifetime. The decisions you make on energy sources and daily conservation practices will determine not only the quality of your life, but also the lives of your descendants.

I cannot stress enough how important it is to gain a good understanding of global warming: what it is, why it is happening, how it can be slowed down, why everybody is contributing to the problem, and why *everybody* needs to be an active part of the solution.

I would sincerely like to thank several of the federal government agencies that research, educate, and actively take part in dealing with the global warming issue—in particular, the National Aeronautics and Space Administration (NASA), the National Oceanic and Atmospheric Administration (NOAA), the Environmental Protection Agency (EPA), and the U.S. Geological Survey (USGS) for providing an abundance of resources and outreach programs on this important subject. I give special thanks to Al Gore for his long and diligent effort toward bringing the global warming issue so powerfully to the public's attention. And I applaud California governor Arnold Schwarzenegger for putting care of the environment high on his agenda. I would also like to acknowledge and give thanks to the many wonderful universities across the United States, in England, Canada, and Australia, as well as private organizations, such as the World Wildlife Fund, that diligently strive to educate others and help toward finding a solution to this very real problem.

I want to give a huge thanks to my agent, Jodie Rhodes, for her assistance, guidance, and efforts, and also to Frank K. Darmstadt, my editor, for all his hard work, dedication, support, and helpful advice and attention to detail. His efforts in bringing this project to life were invaluable. Thanks also to Alexandra Lo Re for her attention to detail and to the copy editing and production departments for their assistance and the outstanding quality of their work.

INTRODUCTION

Hardly any other scientific issue has generated so much controversy as global warming has over the past few decades. Large groups of people strongly agree that global warming is real, it is happening right now, and action must be taken immediately before it is too late for the Earth to recover. Other groups disagree, saying that the present-day warming of the Earth's atmosphere is just part of a natural cycle and cannot be controlled. Still others do not know what to think about all the commotion; whether global warming is a good thing or bad. And it gets even more heated because a large part of this controversy is charged with emotion since it affects everyone on the planet in a personal way.

Although humans may not take the time to think about it every day, the climate we live in affects every aspect of our lives. Every single day, thousands of economic decisions are made based on climatic information. Reliable climate information is necessary in one way or another to make good business decisions in just about every enterprise in America. Construction, agriculture, and aviation rely strongly on climate data. Consumer business, although not as obvious, does as well. For instance, effective marketing strategies target advertising when the potential demand is greatest. Sports stores sell snowboards and skis when it snows and swimsuits when it is warm. Grocery stores sell more popsicles, lemonade, and picnic supplies in the summer. Retail stores sell bug spray, camping gear, and luggage racks during the summer vacation season. Climate also dictates where to build facilities. After all, Disneyland would probably not do as well in Fairbanks, Alaska.

In agriculture and horticulture, climate is extremely important. Seeds need to be planted at certain times, at certain latitudes, and require specific amounts of water to survive. If they do not get this, they will not grow. The type of climate in an area determines what kinds of buildings can be built. In areas that receive excessive amounts of snow, the buildings must be engineered to be able to hold the weight of the snow on their roofs. Related to this, insurance companies care about the climate, because they need to identify potential problem areas that may experience hurricanes, tornadoes, floods, landslides, and mudslides. These areas pay higher insurance premiums. Each one of us must deal with climate every day. What should we wear? Do we need a coat? Sunscreen?

The reality of global warming is that it will cause significant climate change. In fact, many scientists today refer to the phenomenon of global warming as climate change because they feel it is a better overall description of the situation. While it is certainly true that the atmosphere is warming up, that is only one part of what is going on. As the Earth's atmosphere continues to warm, it is setting off an avalanche of other mechanisms, which will do even greater harm to the Earth's natural ecosystems. Glaciers and ice caps are melting, sea levels are rising, and ocean circulation patterns are changing, which then changes the traditional heat distributions around the globe. Seasons are shifting and storms are becoming more intense, leading to severe weather events. Droughts are causing desertification, crops are dying, and disease is spreading. Some ecosystems are shifting where they still can; others are beginning to fail. In short—humans are changing the Earth's climate—and not for the better.

Scientists believe that it is not too late to fix this problem we have brought upon ourselves if we act now. But in order to act, we must first become educated about global warming, so that we know just what we are dealing with. You, the readers, will be the leaders of tomorrow, and you do have the power to do something about this situation.

Climate Systems: Interactive Forces of Global Warming, the first volume in this multivolume set, lays out the basic scientific framework needed to understand how climate systems work and what global warming involves.

Chapter 1 presents the concept of global systems, climate cycles, the atmosphere's structure, and an overview of how the Earth's natural greenhouse effect operates. Chapter 2 discusses the carbon cycle and its link to other major natural Earth cycles, such as biogeochemical cycles and the hydrologic cycle. It also looks at the way carbon is balanced between the land, ocean, and atmosphere and how carbon sequestration and sinks affect that balance.

Chapter 3 examines climate as it relates to the movement of the Earth's continents through the plate tectonic process. It also presents some theories as to how plate tectonics contributes to carbon dioxide levels. Chapter 4 explores the flow of energy. It focuses on solar energy and its key role in dictating the Earth's climate, the Earth's energy, and atmospheric energy. It also explains the various methods by which heat energy is transferred through the atmosphere. It then discusses the concepts of the Earth's overall energy balance and why its maintenance is important to ecosystems.

Chapter 5 presents the planetary and global motions of the atmosphere that affect climate. It explains the varied effects of the Earth's eccentricity, tilt, and precession over time and how these movements affect the climate. It also addresses the Coriolis force—a major global force that is a product of the Earth's rotation that directly influences airflow and storm patterns. It concludes by looking at the large-scale effects of El Niño. Chapter 6 discusses the local motions in the atmosphere that affect weather and climate, from regional wind systems, to local wind systems, to extreme weather and emergency preparedness.

Chapter 7 delves into the Earth's ocean currents and explains ocean circulation, the roles of seawater density, temperature, and salinity, and major ocean currents that directly control climate and how their destabilization could cause an abrupt climate change. Chapter 8 explores the global warming issue itself. It looks at various scientific viewpoints on the issue, how advances in technology and education have influenced it, the collective growth of environmental awareness in the United States, and the effect that public and media response have had on the issue.

Chapter 9 compares scientific inquiry to the limits of technology and explains how they differ. It discusses the concepts of positive and negative feedback, the mission of the Intergovernmental Panel on Climate Change, and what is known and not currently known about global warming. Chapter 10 looks at the consensus about global warming from the majority of scientists, which countries are major contributors to the problem, strategies for coping with global climate change, current research, and what lies ahead.

My goal as author is to open the doors to this incredibly controversial topic and to challenge you to research the relevant issues further and decide for yourselves what kind of an environment you want to see in the future, for you, your children, and your grandchildren.

Elements of the Climate System

Energy from the Sun drives the Earth's weather and climate. The atmospheric gases are nitrogen, oxygen, and the trace gases, including the noble gases and the greenhouse gases (water vapor, methane, carbon dioxide, ozone, and nitrous oxide). These greenhouse gases trap some of the energy from the Sun, creating a natural greenhouse effect, without which normal temperatures would be much lower and the Earth would be too cold to live on. Thanks to the greenhouse effect, the Earth's average temperature is 60°F (16°C). Problems begin when the natural greenhouse effect is enhanced by human-generated emissions of greenhouse gases. Scientists now predict that the Earth's climate will change because human activities are altering the chemical composition of the atmosphere through the buildup of greenhouse gases. In order to understand global warming, it is necessary to understand climate. This chapter introduces the elements of the climate system. It distin-

1

guishes climate from weather, introduces the concept of global systems, provides an overview of the atmosphere's physical structure, and introduces the greenhouse effect and global warming.

CLIMATE AND WEATHER

The distinguishing factors between weather and climate are (1) the time interval in which they are taking place (such as a day versus a season) and (2) the scale of the area they are taking place over (such as a village versus a country). Weather is what the atmosphere is currently doing— snowing, raining, or clear skies. Climate is how the atmosphere behaves over relatively long time intervals—such as hot summers, cool winters, or wet springs.

Climate also refers to the average condition of a region, measured in characteristics such as temperature, amount of rainfall or snowfall, how much snow and ice cover there is on the ground, and the characteristics of the predominant winds; the long-term trends of an area. As an illustration, Hawaii may be described as having a tropical climate, Britain a maritime climate, and Saudi Arabia an arid climate. Climate applies to long-term changes (months, years, and longer); weather the shorter fluctuations that last hours, days, or weeks. The weather may be snowing, raining, clear and sunny, hot, or windy. Weather conditions generally apply only to limited geographical regions and can change rapidly.

When scientists say that climate has changed, they are referring to the fact that the averages of the daily weather have changed over time—such as a region is drier, receiving half as much rain, than it was 50 years ago. There can also be short-term climatic variations, such as those associated with El Niño, La Niña, or volcanic eruptions. These will be looked at in greater detail in chapter 5.

Climate affects everyone on Earth in some way, just as global warming will affect everyone on Earth. The weather plays some part in everything humans do, whether it is in transportation, growing food, producing an adequate water supply, or producing and manufacturing goods. Global warming will have an impact on every human on the globe, in varying intensities. Scientists have determined that rising global temperatures will change precipitation patterns, melt ice caps, raise sea levels, affect water supplies, damage the world's forests, spread disease, cause both floods and droughts, and encourage hazardous weather events.

All of the Earth's ecosystems will be affected. Because crop yields, food systems, and water supplies are being jeopardized as the Earth's atmosphere heats up at an accelerated rate, it is critical that scientists study, understand, and unlock the mysteries of global warming now before more damage is done. Technology has advanced to the point that climatologists (scientists who study the climate) have developed several tools that allow them to not only study current climate, but piece together evidence of past climates, and build models to predict future climates.

It is important to understand that even without global warming, the Earth's climate is always fluctuating. Change is natural. The Earth's climate has changed throughout geologic time as the Earth's continents have shifted positions, as the Earth's orbit and the tilt of its axis have changed, and even as the chemical composition of the atmosphere has evolved.

The Earth's climate system consists of air, land, water, ice, and vegetation. Climatologists study these components in terms of their cause and effect—also called forcing and response. The term *forcing* describes the things that cause the change; the *responses* are the changes that occur. Figure 1 (on page 4) illustrates the overall climate system along with the interaction of the components within the system. Sometimes climatologists find that the forcings can be cyclic in nature. For instance, various changes in the Earth's orbit are cyclic, and this pattern is universally reflected in climate data.

In order to fully understand climate, scientists must study climatic records that go back thousands to millions of years. Unfortunately, historical weather records (such as rainfall, sea level, and temperature) have only been kept for the past 100 years or so. In order to infer climate change of the distant past, a *proxy* (an object whose physical properties represent climatic conditions of the past, such as tree rings, ice cores, lake sediments, and corals) must be used.

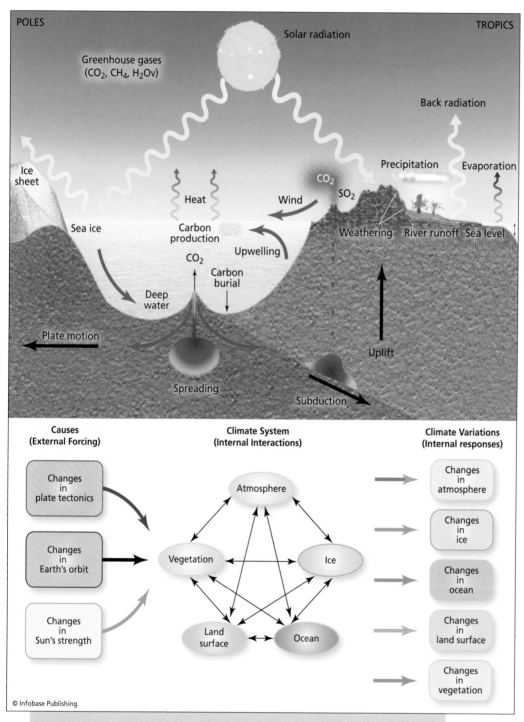

Earth's climate system is composed of many components that all interact. A change in any one component causes changes in others.

FEDERAL AGENCIES STUDYING CLIMATE

The Climate Prediction Center (CPC) was established in the 1980s by the U.S. National Weather Service. It is the agency that forecasts the El Niño and La Niña events in the tropical Pacific. It provides forecasts of climate change as well as real-time monitoring of climatic events by monitoring the atmosphere, the land, and the ocean, supplying vital data for industries, such as transportation, energy, health, water resources, and agriculture.

The National Aeronautics and Space Administration (NASA) is also heavily involved in global warming and climate change research. Well known for their use of satellite and computer modeling technology, they have developed sophisticated models that are able to calculate specific surface temperatures (solar radiation reflected and absorbed) around the world and measure how much they have been warming.

NASA's satellites also monitor other important aspects related to climate and global warming, such as volcanic eruptions, melting ice sheets and glaciers, El Niño, and changes in global wind and pressure systems. This information can be fed into their supercomputers to model scenarios of global warming.

THE CONCEPT OF A GLOBAL SYSTEM

Although different types of climate exist in different parts of the world, the climate does work as a complete global system. Conditions and actions in one area influence conditions and actions in other areas. Because the Earth shares one atmosphere, what goes on, for example, in China will affect the United States.

Thus, global warming is also a global issue: It is affecting the Earth in different ways, such as destabilizing major ice sheets, melting the world's glaciers, raising sea levels, contributing to extreme weather, and shifting biological species northward and higher in elevation. No area of the Earth is immune from its effects. Scientists at NASA have determined that the atmospheric concentration of carbon dioxide (an important greenhouse gas) is now higher than it has been for the past 650,000 years.

Both the circulation of the atmosphere and the oceans occur on a global scale. Major currents in the ocean carry huge amounts of heat from the equator to the poles. There are global ocean currents that circulate heat energy on the surface and at great depths, connecting the Earth's major oceans. One extremely important current moves in a winding, endless loop; scientists refer to its conveyor belt–like properties as the thermohaline circulation (THC). This current is significant to major parts of the world—it moves the warm salty Atlantic water that originates near the equator northward toward Greenland and Labrador, where it then cools and sinks. The current sinks more than one mile in very specific places, where it then flips around, heads south, and makes its way back through the Atlantic toward the equator again. From there, the water continues to move south, travels around the southern tip of Africa, and rises to the surface in the Indian and Pacific Oceans, as well as areas near Antarctica. It then heads north toward the equator again, where it picks up heat, and repeats the cycle. With the loss of this current, western Europe's winters would get much colder.

The loss of this huge circulation system would have other unforeseen effects as well. Global warming must be approached as a global issue. It will take everyone working toward a solution to make a difference.

CLIMATE VARIABILITY AND CYCLES

Climate has not remained the same throughout time. Many natural phenomena have had their effects on the Earth's climate on different timescales: on tectonic scales are the slow movements of the Earth's continents—a process called plate tectonics (discussed in greater detail in chapter 3), on orbital scales are the orbital variations of the Earth, its eccentricity, tilt, and wobble (discussed in chapter 5); on millennial scales, ice ages, and on centennial scales, droughts.

Because these natural cycles can be occurring independently and take different amounts of time to occur—tectonic scales have occupied periods over the past 300 million years; orbital scales over the past 3 million years; millennial scales over the past 50,000 years, and centennial scales over the past 1,000 years—the climate system can be very complex because all these climate processes interact. In general, the shorter the timescale, the more impact a process has on a local scale

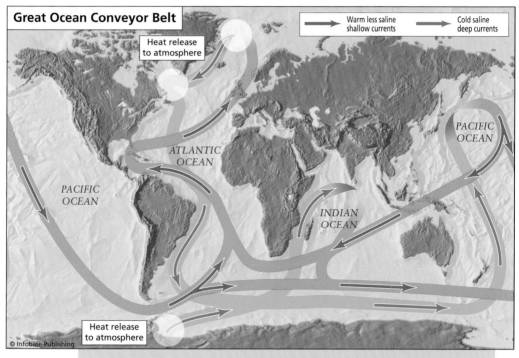

The thermohaline circulation, or ocean conveyor belt, is the principal mechanism for transporting heat from the Tropics to the polar regions. If the conveyor belt were to shut down, it could cause Europe to enter a severe cold period and disrupt climate around the world. *(modeled after IPCC)*

(such as a drought). The longer the timescale, the bigger impact the process has on a global scale (such as plate tectonics).

WHAT EXACTLY IS GLOBAL WARMING?

Global warming is a term scientists use to refer to the increase of the Earth's average surface temperature, due largely to a buildup of greenhouse gases in the Earth's atmosphere. The term was first used in a paper, "On the Influence of Carbonic Acid in the Air upon the Temperature of the Ground" published in 1896 by Svante Arrhenius, a Swedish scientist and Nobel laureate. Often, scientists will refer to the problem as climate change to convey the concept that there is actually more going on than just rising temperatures. Climate change encompasses long-term changes in climate, which include temperature, precipitation amounts, and types of precipitation, humidity, and other factors.

Today, global warming is one of the most controversial issues in the public eye. It is discussed frequently in print and on televised news reports, in documentaries, scientific and political debates, classrooms,

GLOBAL WARMING: MYTHS AND FACTS

There are several myths circulating about global warming. A few of them are listed below, followed by their corresponding facts.

Myth: Global warming is too uncertain. We do not really know if, or why, the atmosphere is changing.

Fact: There is now overwhelming scientific evidence that global warming is occurring. There are not only many physical indicators that scientists can measure today, but also indicators exist in records from ancient times that show us how the environment has changed. Thousands of scientists overwhelmingly support this evidence and believe that the warming trend will worsen and that current human activity is largely to blame.

Myth: Even if the Earth is warming, maybe that will be a good thing.

Fact: In the short term, there may be some areas that benefit from global warming. For instance, some areas may now be able to be farmed at some latitudes that were previously too cold, but most likely will not be very productive. In the long term, any possible benefits from global warming will be very small compared to the extreme negative effects worldwide.

Myth: Even if global warming is real, doing something about it will hurt the U.S. economy.

Fact: Companies worldwide that are already reducing their carbon emissions are finding that cutting pollution can be economically beneficial. For example, utility companies switching to wind power are creating new jobs, boosting their economies. Using skills and ingenuity can start new industries geared toward carbon-free technology and production. Even the world's major oil companies are currently getting involved in developing renewable energy resources.

and other venues. It receives a lot of attention because it is more than a scientific issue—it also affects economics, sociology, and people's lifestyles and standards of living. It is one of the most passionate political issues, not only in the United States, but worldwide as public demands for a solution have intensified.

Myth: The Earth has warmed up before. Maybe what is happening today is natural.

Fact: The global warming of today is not natural. Scientists have already considered—and ruled out—natural explanations. Today's CO_2 levels are the highest they have been in the past 650,000 years.

Myth: Civilization can adapt to climate, whether it is hot or cold.

Fact: Changes in climate have been responsible for the destruction of many civilizations in the past. At the very least, global warming will cause major hardships and suffering, both physically and financially. Even worse, it will affect not just those living today, but for generations to come. Scientists also warn that it is expected there will be a greater warming trend than human civilization has ever faced in the past 10,000 years.

Myth: If the ozone hole shrinks, global warming will go away.

Fact: The ozone hole and global warming are two different problems. Global warming involves the lower part of the atmosphere (the troposphere) and is caused by the increasing concentrations of heat-trapping gases. The ozone hole involves the loss of ozone in the upper part of the atmosphere (the stratosphere), which allows incoming harmful ultraviolet radiation from the Sun to reach the Earth's surface.

Myth: Global warming is not happening because not all of the glaciers and ice sheets are melting.

Fact: Most of the world's glaciers have been retreating. Some have had regional gains where storm frequency has increased. The overall trends are that ice sheets are melting and many of the rates are accelerating.

More than 2,500 of the world's most renowned scientists from many diverse disciplines, represented by the United Nations Intergovernmental Panel on Climate Change (IPCC), support the concept of global warming and agree that there is absolutely no scientific doubt that the atmosphere is warming. They also believe that human activities—especially burning fossil fuels (oil, gas, and coal), deforestation, and environmentally unfriendly farming practices—are playing a significant role in the problem.

The science of global warming does not come with a crystal ball. Scientists do not know exactly what will happen, such as what the specific impacts to specific areas will be, nor can they say with certainty when or where the impacts will hit the hardest. But they are certain that the effects will be serious and globally far-reaching. According to the NOAA/NASA/EPA Climate Change Partnership (National Oceanic and Atmospheric Administration, National Aeronautics and Space Administration, and Environmental Protection Agency), potential effects include increased human mortality, extinction of plant and animal species, increased severe weather, drought, and dangerous rises in sea levels.

Although climatologists still argue about how fast the Earth is warming and how much it will ultimately warm, they do agree that global warming is happening right now and that the Earth will continue to warm if something is not done soon to stop it.

THE ATMOSPHERE'S STRUCTURE

The atmosphere can be thought of as a thin layer of gases that surround the Earth. The two major elements—nitrogen and oxygen—make up 99 percent of the volume of the atmosphere. The remaining 1 percent is composed of what are referred to as trace gases. The trace gases include water vapor, methane, argon, carbon dioxide, and ozone. Although they only make up a small portion of the atmosphere, the trace gases are very important. Water vapor in the atmosphere is variable: Arid regions may have less than 1 percent, the Tropics may have 3 percent, and over the ocean there may be 4 percent.

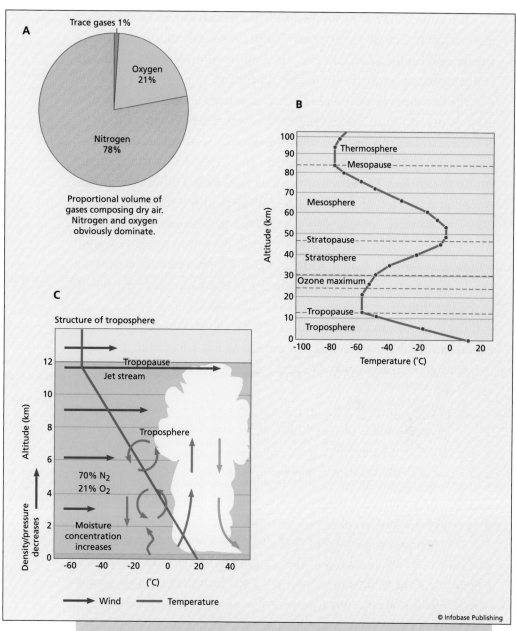

A Trace gases 1%

Oxygen 21%

Nitrogen 78%

Proportional volume of gases composing dry air. Nitrogen and oxygen obviously dominate.

B

Thermosphere
Mesopause
Mesosphere
Stratopause
Stratosphere
Ozone maximum
Tropopause
Troposphere

Altitude (km)
Temperature (˚C)

C

Structure of troposphere

Tropopause
Jet stream
Troposphere
70% N_2
21% O_2
Moisture concentration increases

Altitude (km)
Density/pressure decreases
(˚C)

→ Wind   — Temperature

© Infobase Publishing

The Earth's atmosphere is composed of 78 percent nitrogen, 21 percent oxygen, and 1 percent trace gases; the atmosphere is divided into four layers based on temperature; the troposphere is the layer in which all the weather occurs. It averages seven miles (11 km) high.

The atmosphere is not uniform. It is divided vertically into four distinct layers: (1) the troposphere, (2) the stratosphere, (3) the mesosphere, and (4) the thermosphere. The four levels are divided by height and temperature, as shown in the illustration on page 11.

The lowest level, closest to the Earth's surface is the troposphere. It extends upward to an average height of seven miles (11 km). This level is critical to humans because all of the Earth's weather occurs in this layer. In this level, the temperature gets cooler with increasing height. At Earth's surface the temperature averages 59°F (15°C) and at the tropopause (the top of the troposphere) the temperature is only -71°F (-57°C). Moisture content also decreases with altitude in this layer. Above the troposphere there is not enough oxygen to sustain life and winds increase with height. One of the most pronounced wind systems—the jet stream—is located at the top of the troposphere, as shown in the illustration.

The climate of an area is the result of both natural and anthropogenic (influenced by humans) factors. The natural factors come from the

- Atmosphere (air)
- Lithosphere (land)
- Biosphere (life)
- Hydrosphere (water)

The human factor influences climate when it alters land and resource uses. For example, when people change a natural forested area to a city, it has a direct effect on climate.

When climate change occurs naturally, it is so slow (over thousands of years) that it is not readily detectable. Climate changes caused by people occur much faster and are noticeable within a few generations, or less. Often, a change in one part of the climate produces changes in other parts as well, because the Earth is a global system.

Since the Industrial Revolution began in the late 18th century—and especially since the introduction and use of fossil fuels involved in the rapid modernization of the 20th and 21st centuries—the global average temperature and atmospheric carbon dioxide (CO_2) concentrations have increased notably. Because CO_2 levels are higher now than they have been in the past 650,000 years and surface temperatures on Earth have risen significantly during the same time, scientists have concluded that humans are responsible. In support of this, scientists at both NOAA and NASA have run two types of computer models: one of climate systems with natural climate processes alone and another with natural climate processes combined with human activities. The models that include the human activities more accurately resemble the actual climate measurements of the 20th century, giving scientists further proof that human activity does play a significant role in global warming.

Since the 1970s, the Earth's climate has been monitored by instruments on satellites. In addition, measurements of the atmospheric CO_2 have been obtained since 1957, when the world's first monitoring station was built on top of Mauna Loa—the highest mountain on Hawaii and the Earth's largest volcano. Scientists wanted to monitor the atmospheric CO_2 levels to check if they were increasing. They were able to determine a definite increase—one that is still occurring today. Currently, the CO_2-monitoring network has expanded to more than 100 stations globally in order to track concentrations of carbon dioxide, methane, and other greenhouse gases.

Scientists know natural climate variability and cycles will continue, and they expect CO_2 levels to rise and global warming to increase because of human influences. How much climate change occurs will ultimately depend on the choices humans make concerning population, energy, technology, and global cooperation.

Dr. Rajendra Pachauri, chairman of the IPCC, said at an international conference attended by 114 governments in Mauritius in January 2005, that the world has "already reached the level of dangerous concentrations of carbon dioxide in the atmosphere." He recommended immediate and "very deep" cuts in the pollution levels if humanity is to "survive." He also said: "Climate change is for real. We have just a small window of opportunity and it is closing rather rapidly. There is not a moment to lose."

In a report in the *Guardian* in February 2006, Dave Stainforth, a climate modeler at Oxford University, said: "This is something of a hot topic but it comes down to what you think is a small chance—even if there's just a half percent chance of destruction of society, I would class that as a very big risk."

The world's first functioning CO_2-monitoring station is on top of Mauna Loa—the highest mountain on Hawaii and the Earth's largest volcano. Here scientists are able to monitor atmospheric CO_2 levels. This facility has been instrumental in providing well-documented evidence that CO_2 levels are steadily rising—evidence supporting the existence of global warming. *(NOAA)*

Chris Rapley, head of the British Antarctic Survey, commented in 2006 that the huge West Antarctic ice sheet may be starting to disintegrate, an event that would raise sea levels around the world by 16 feet (5 m). He said, "The IPCC report characterized Antarctica as a slumbering giant in terms of climate change. I would say it is now an awakened giant. There is real concern."

According to the American Geophysical Union, "Natural influences cannot explain the rapid increase in the global near-surface temperatures observed during the second half of the 20th century."

THE GREENHOUSE EFFECT AND GLOBAL WARMING

In order to understand global warming, it is necessary to understand what the greenhouse effect is. In order to understand the greenhouse

THE MAUNA LOA OBSERVATORY

Scientists first began to suspect that fossil fuels were contributing to global warming in the mid-1950s. Roger Revelle, a scientist at the Scripps Institution of Oceanography, used the metaphor that humans were using the Earth's atmosphere to carry out a "giant experiment"—one significant enough to potentially alter the Earth's climate.

As scientists became interested in studying CO_2 and the ways in which it interacts with the atmosphere, the biosphere, the lithosphere, and the oceans, a working atmospheric observatory was set up on the top of Mauna Loa, Hawaii, by the United States Weather Bureau in 1956. Continuous CO_2-monitoring was begun.

Dr. Charles "Dave" Keeling of the California Institute of Technology also played a significant role in setting up the continuous monitoring program. He felt Mauna Loa was an ideal location for the observatory because it was a natural source of clean air—in the middle of the Pacific Ocean away from sources of pollution.

Keeling kept close records year after year of the CO_2 levels. He plotted the readings he developed from these records, which resulted in the Keeling Curve, a very distinctive upward trend of CO_2 levels from 1958 to the present. The Keeling Curve is considered "the cornerstone of global warming science." In 1997, Keeling was honored at a White House ceremony by Vice President Al Gore with a special achievement award "for forty years of outstanding scientific research associated with monitoring atmospheric carbon dioxide in connection with the Mauna Loa Observatory." In 2002, President George W. Bush selected Keeling to receive the National Medal of Science, the nation's highest award for lifetime achievement in scientific research.

Today, the Mauna Loa Observatory is recognized as the world's premier long-term atmospheric monitoring facility and has the distinction of being the site where the continually increasing concentrations of global atmospheric CO_2 were first discovered. Here, it was confirmed that the mean atmospheric concentration of CO_2 was 316 parts per million by volume (ppmv) in 1958, rose to 369 ppmv by 1998, 379 ppmv by 2005, and has increased every year to the present. By the end of 2007, it had reached 383 ppmv.

(continues)

(continued)

The Mauna Loa Carbon Dioxide Record is the longest continuous record of atmospheric concentrations of CO_2. According to NOAA, it is often called the "most important geophysical record on Earth" and has been instrumental in showing that mankind is indeed changing the composition of the atmosphere through the combustion of fossil fuels. This is what formed the basis for the theory of global atmospheric change through heating of the Earth's atmosphere.

Currently, data from the observatory is used by climate scientists and modelers from all around the world to model and predict what the Earth's future climate may be like. Every year, the Mauna Loa Observatory releases new data, which is used by scientists in models, politicians in international plans and agreements, and managers in business, production, and policy making.

effect, it is necessary to understand the properties of the Sun's radiation and the way the radiation interacts with the Earth's atmosphere, oceans, land surfaces, and what is on the ground, such as vegetation, lakes, and buildings.

It is the Sun's energy that drives the Earth's weather and climate. The Sun radiates enormous amounts of energy throughout space in a wide spectrum of wavelengths, ranging from very short (high energy) to very long (lower energy). This is called the electromagnetic spectrum. Humans can only see a small portion of the electromagnetic spectrum, which is referred to as visible light and can be broken down into the colors of the rainbow, with purple the shortest wavelength to red as the longest, as shown in the illustration on page 17.

The narrow band of visible light represents 43 percent of the total radiant energy emitted from the Sun. When the Sun's energy reaches the Earth, it interacts in a multitude of ways with the variety of surfaces it comes in contact with. It can be reflected, absorbed, conducted,

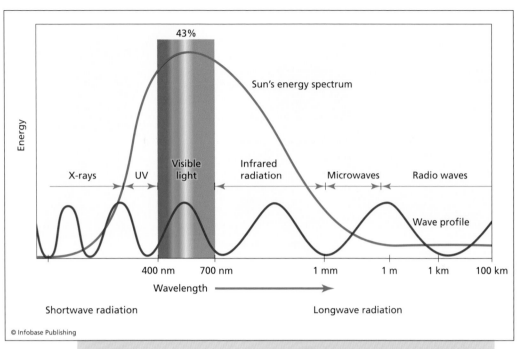

The Sun's electromagnetic spectrum ranges from short wavelengths, such as X-rays, to long wavelengths, such as radio waves. The majority of the Sun's energy is concentrated in the visible and near-visible portion of the spectrum—the wavelengths located between 400 to 700 nanometers (nm).

stored, convected, and reradiated. It is because of the special composition of the Earth's atmosphere that the incoming heat energy from the Sun does not just strike the Earth, bounce back off, and get reradiated back into space. The visible light is transformed into heat and reradiates in the form of invisible infrared radiation (the energy changes to longer wavelengths that humans cannot see and becomes heat).

Albedo and particulate matter in the atmosphere also affect the incoming solar radiation. Albedo refers to the reflectivity of a surface. A highly reflective surface—such as ice or snow—reflects most of the radiation that strikes it. A surface with a low albedo absorbs most of the radiation. A dark surface has a low albedo, so it absorbs much of the radiation striking it and heats up.

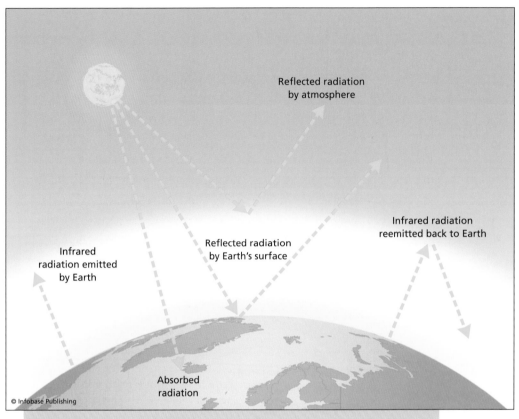

Reflected radiation
by atmosphere

Infrared radiation
reemitted back to Earth

Infrared
radiation emitted
by Earth

Reflected radiation
by Earth's surface

Absorbed
radiation

© Infobase Publishing

When incoming solar energy reaches the Earth, it interacts with the atmosphere and the surface of the Earth in many ways. It can be scattered, absorbed, or reflected by clouds, particles in the atmosphere, and the surface of the Earth. The atmosphere can also reradiate energy back to the Earth's surface.

If there is a lot of small particulate matter—such as dust and smoke—in the atmosphere it can scatter radiation, keeping it from being absorbed. This phenomenon is what creates hazy skies. One situation, called *global dimming*, happens when there is so much particulate matter in the atmosphere that it screens out incoming radiation, causing a cooling effect. According to scientists at NASA, global dimming may counteract some of the effects of global warming. This means that as humans reduce pollution and its accompanying particulate matter it may actually accelerate global warming.

If the atmosphere were transparent and could not retain any heat, then each day the Earth would warm up and at night all the accumulated energy would reradiate back into space and the Earth's temperature would quickly drop below 0°F (-18°C).

As noted earlier, less than 1 percent of the Earth's atmosphere is composed of greenhouse gases. It is these greenhouse gases (water vapor, CO_2, methane [CH_4], halocarbons [HFCs, CFCs, HCFCs], ozone [O_3], and nitrous oxide [N_2O]) that absorb the heat radiated from the Earth's surface, trapping the heat and radiating it back to the Earth's surface. This is what is commonly referred to as the greenhouse effect. It is likened to a greenhouse because it works in much the same way as a gardener's greenhouse does. In a real greenhouse, the glass panels let in light, which heats the inside and keeps the plants, soil, and interior of the building warmed. The greenhouse traps the heat, keeping it from escaping. That is why the temperature is always much warmer inside a greenhouse.

A greenhouse maintains a warm environment for plants because sunlight enters through its glass panels and heats up the inside. It then traps the heat, keeping the interior warm. *(Nature's Images)*

Only Earth can support life, due largely to the natural greenhouse effect of the Earth's atmosphere. *(Lunar and Planetary Institute, NASA)*

The Earth's atmosphere and its heat-retaining capabilities are critical to life on Earth. Without the greenhouse effect, the Earth would be a very cold planet; its surface temperature would be well below freezing. The greenhouse effect serves to insulate the Earth, creating the mild temperatures that make life possible. When comparing life on Earth to possibilities of life on the neighboring planets of Venus or Mars, some scientists refer to the Earth's greenhouse effect as the Goldilocks principle: It is not too hot, not too cold, but just right. The Earth has a comfortable average surface temperature between the boiling and freezing points of water, and distance from the Sun is only part of the reason why this is so. The rest is due to the composition of the Earth's atmosphere. When Earth is compared to its solar system neighbors, Venus has a thick atmosphere of CO_2, making it much too hot to support life—its surface temperature is about 860°F (441°C). Mars, on the other side of the spectrum, has much too thin a layer of CO_2 and is too cold to support life—it averages -81°F (-63°C).

Changes in the composition of the Earth's atmosphere can alter the greenhouse effect. This has happened naturally throughout the Earth's climate history, with cycles of cooler and warmer periods. Today, however, humans are having a tremendous effect on some of the main elements that determine climate by changing the composition of the atmosphere and changing the Earth's surface. According to the National Academy of Sciences, the Earth's surface temperature has risen by about 1°F (0.6°C) in the past 100 years. Some places have seen a temperature increase of equal to or more than 3.3°F (2°C). Such places include Siberia, which has warmed 3.3°F (2°C) since 1970 (three times the global rate), the Alaskan interior and northern Canada, which have warmed 3.3°F (2°C) since 1950; the Alaskan permafrost areas, which have warmed 4.2°F (2.5°C); and the Arctic Sea. The polar and other high latitude areas are the most susceptible for accelerated temperature rise. The reason for the faster warming in the polar regions is due to the complex feedback mechanisms that exist between the atmosphere, the ocean, and ice. White snow and ice reflect most of the energy coming from the Sun while dark oceans and land absorb this energy. As snow and ice cover melt because of global warming, an increasing amount of energy is absorbed, and the resulting warming becomes self-perpetuating through a positive feedback mechanism. As the system becomes unbalanced, it gets out of control and more difficult for scientists to model and predict. Today, scientists at the Norwegian Polar Institute claim that these processes have been advancing at a faster pace than most researchers have predicted. They state that the warming of the Arctic is much more pronounced than the rest of the Earth's ecosystems.

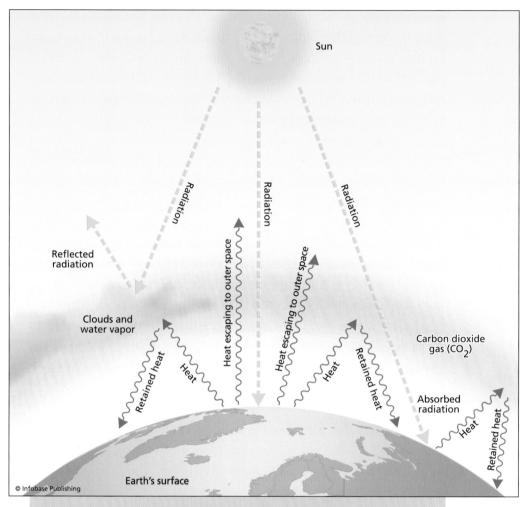

Sun

Radiation

Radiation

Radiation

Reflected
radiation

Heat escaping to outer space

Heat escaping to outer space

Clouds and
water vapor

Retained heat

Heat

Heat

Heat

Retained heat

Carbon dioxide
gas (CO$_2$)

Absorbed
radiation

Heat

Retained heat

Earth's surface

© Infobase Publishing

The enhanced greenhouse effect due to human influences has increased the Earth's temperatures since the beginning of the Industrial
Revolution, largely through the burning of fossil fuels, industrialization, and deforestation.

Greenhouse gases constantly absorb and emit heat energy, so the
Earth's surface gets heat from both the Sun and gases. When greenhouse gases absorb some of the outgoing radiation, they reradiate infrared radiation in all directions, increasing temperatures. Any energy
trapped between the Earth and atmosphere heats up the Earth.

One of the biggest causes of increased greenhouse gases has been
the use of fossil fuels (coal, oil, and natural gas). Since the Industrial

Revolution began, human activities have greatly increased the addition of CO_2 to the atmosphere. The actual values are alarming. The CO_2 concentration in the atmosphere has risen 30 percent since the late 1800s. The preindustrial atmospheric CO_2 level was about 270 parts per million; the current level is 370 parts per million. In their 2007 report, the IPCC projects that if humans do not cut back current industrial and lifestyle practices, CO_2 concentrations in the atmosphere could reach 600 to 970 parts per million by the year 2100. This means that the average global temperature would probably rise by 2.7 to 10.4°F (1.4–5.8°C) between 1990 and 2100.

Another activity that has contributed greatly to the rise in CO_2 is the destruction of forests worldwide in order to change use of the land to agriculture, cities, and other human-oriented uses. Forests store huge amounts of CO_2 (all vegetation stores CO_2), so when forests are removed, so also is the reservoir of CO_2. As an illustration, the world's tropical forests hold an enormous amount of carbon. The plants and soil in the Earth's tropical forests hold roughly 507 to 634 billion tons (460–575 billion metric tons) of carbon. Divided, this means that each acre of tropical forest stores about 198 tons (180 metric tons) of carbon. Not only will destroying the forest prevent future CO_2 storage, but also if the existing forest is burned, the CO_2 that has been stored away for years is suddenly released back into the atmosphere. When a forest is cut down, the carbon that was stored in the tree trunks joins with the oxygen in the atmosphere and produces CO_2. This has a tremendous effect on the global carbon cycle. In fact, from 1850 to 1990, deforestation worldwide released 134 billion tons (122 billion metric tons) of carbon into the atmosphere. In comparison, all of the fossil fuels burned each year equal roughly 7 billion tons (6 billion metric tons) per year.

Another huge complication for the planet is increasing population size. As increased stress is put on the Earth's resources, as more land is urbanized and more fossil fuels burned, scientists warn that global warming will worsen if positive action is not taken now. Human interaction with the natural greenhouse effect is called the enhanced greenhouse effect, and it is this enhanced greenhouse effect that has earned a negative reputation by those concerned about the health of the environment. The table below illustrates global population growth from 1800, projected to the year 2050.

Scientists at NASA are convinced that human influence is responsible for the sudden increase in recent temperature and CO_2 levels. They have already documented the following changes in the Earth's climate:

- Global average surface temperature has increased more than 1°F during the 20th century alone
- The 18 warmest years of the 1900s occurred after 1980
- The 1998 global temperature set a new record
- Higher latitudes (polar regions) have warmed more than lower latitudes (tropical regions)
- Many of the world's glaciers and ice caps are melting
- Nighttime temperatures have risen more than daytime temperatures

Global Population 1800 to 2050	
YEAR	POPULATION
1800	1 billion
1922	2 billion
1959	3 billion
1974	4 billion
1987	5 billion
1999	6 billion
2013	7 billion
2028	8 billion
2050	9 billion

Source: United Nations; U.S. Census Bureau, International Programs Center, International Data Base and unpublished tables.

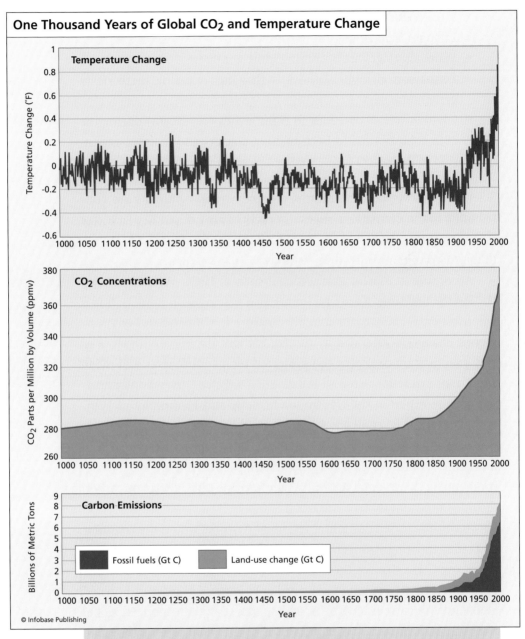

One Thousand Years of Global CO$_2$ and Temperature Change

Temperature Change

Temperature Change (°F)

Year

CO$_2$ Concentrations

CO$_2$ Parts per Million by Volume (ppmv)

Year

Carbon Emissions

Billions of Metric Tons

Fossil fuels (Gt C)   Land-use change (Gt C)

Year

© Infobase Publishing

The record of the past 1,000 years dealing with temperature change and CO$_2$ concentrations when compared with fossil fuel use and land use change illustrates the dramatic increase in CO$_2$ levels since the Industrial Revolution.

- More water evaporates from lakes, rivers, and oceans
- Drought is becoming more widespread
- Water shortages are becoming more common
- Permafrost at high latitudes is thawing, releasing methane (a greenhouse gas)
- Global sea level rose four to eight inches (10–20 cm) during the 20th century (due to both melting glaciers and thermal expansion of the water)

Scientists believe these conditions are not natural—humans are contributing to the dramatic changes. If practices are not put in place now to curtail them, the world we leave to future generations will be very different from the world we enjoy today.

According to Martin Parry, cochairman of IPCC Working Group II, "For the first time, we are no longer arm-waving with models; this is empirical data, and we can actually measure it. Climatic and environmental changes accelerated by human activity already are under way, and scientists are confident the projected temperature rises will mean significant extinctions around the globe." Dr. Peter Cox, who is a leading climate expert at the Centre for Ecology and Hydrology in Winfrith, Dorset, believes all governments need to listen to what scientists are warning about concerning the future effects of global warming. If humans continue to burn fossil fuels and maintain their other environmentally detrimental actions at the present rates, levels of carbon dioxide in the atmosphere will reach 550 ppm by around 2050.

According to Senator John McCain of Arizona, who is the coauthor of a bill mandating stronger action, "the argument about global warming is over." He stresses that humans have contributed largely to the problem and assertive action is needed now. James E. Hansen, renowned global warming expert at NASA, also speaks of human-induced global warming and climate change and warns of the ramifications if human actions are not slowed immediately. He warns that if the present environmental policies continue unchecked, by the year 2100 the Earth will be a completely different planet.

Dr. Michael Mann, director of the Earth System Science Center at Pennsylvania State University, is one of the leading authorities on global climate change. His research has been focused on establishing the growing human influence on climate. He cites evidence obtained from instrumental measurements taken from proxies reflecting the past 200 years—a branch of climatology called paleoclimatology. Using past evidence found in tree rings he has been able to positively establish the growing negative human influence on climate.

Finally, Dr. Rajendra Pachauri, the chairman of the IPCC, said "The world has already reached the level of dangerous concentrations of carbon dioxide in the atmosphere and there needs to be very deep cuts in the pollution if humanity is to survive."

It is going to take the efforts of the global community to work together to stop the global warming that is changing the climate. Massive public education campaigns, responsibility from elected officials, lifestyle changes for everyone, and laws that will be enforced for noncompliers are some of the programs that must be introduced in order to successfully solve the problem.

Instead of ignoring the efforts of the United Nations, as has been the norm in the United States, responsible officials and the media should be stressing the IPCC's conclusions. For example, Dr. Rajendra Pachauri, chairman of the IPCC (which in 2007 won the Nobel Peace Prize), has recently suggested (September 2008) that all people should have one meat-free day a week if they want to make an effective sacrifice to tackle climate change.

The Union of Concerned Scientists believes global warming is a challenge that every person on Earth can take on and meet, but that action must begin now. Procrastination is not an option anymore. They stress commonsense solutions: fuel-efficient vehicles, renewable energy, and protection of threatened forests. They state that not only will these commonsense solutions reduce global warming, but they also will save people money and create new business opportunities. Best of all, the solutions exist right now. People just need to insist that business and governmental organizations take appropriate steps to ensure that solutions are made both available and affordable to the public. Consumers need to be made aware which choices are best, and those solutions must be made available.

The Carbon Cycle and Its Links to Other Major Cycles

Carbon (C) is a chemical element found in group 14 in the periodic table. It is abundant on Earth and exists in more than 1 million compounds—including diamonds, gas, coal, rocks, shells, and many other things. In fact, all of the living matter on Earth is composed of carbon—all plants, animals, and humans. Carbon is a critical building block of life. This is why archaeologists and paleontologists use carbon in dating techniques when they are trying to determine the age of a very old object—a technique called radiocarbon dating, or C-14 dating.

When anything containing carbon is burned, the carbon reacts chemically with oxygen (O) in the air and creates a gas called carbon dioxide (CO_2), which is released into the atmosphere. Carbon dioxide is one of the most common greenhouse gases and is a major contributor to global warming. It also remains in the atmosphere for a long time. Thus, the CO_2 that is causing global warming today was released into the atmosphere decades ago. At the same time, the CO_2 that humans are emitting today will still be affecting generations long in the future, which is why cutting back on CO_2 should be important to everyone. The good news is that scientists have done research toward identifying the sources of CO_2 emissions, and there now exist technologies that are able to lower these emissions. This chapter examines the natural carbon cycle and how humans are changing it, the hydrologic cycle and the impact global warming is having on it, and how major cycles and systems on Earth are balanced.

BIOGEOCHEMICAL CYCLES

The Earth is a living active planet, always changing, and functions through various energy and chemical cycles. There are continual interactions between the biosphere (life), lithosphere (land), hydrosphere (water), and atmosphere (air) during these cycles. Various substances on Earth move endlessly throughout these four spheres. Of the four spheres, the atmosphere transports elements the fastest. Water, for example, evaporates into the atmosphere from both the land and the ocean, where it condenses and falls back to the land and oceans, where it is then used by plants, animals, and people, fills rivers, reservoirs, and cycles through lakes. During the lifetime of a water molecule, it changes states (liquid, solid, or vapor) many times and can alternately exist in the form of rain, snow, ice, steam, sleet, hail, and water vapor. Over time, that same water molecule may have occurred in a blizzard, a glacier, a tornado, a hurricane, a tsunami, a flood, a simple spring shower, or a glass of water at a dinner table.

The substances may exist in different forms depending on which sphere they are in, such as water can be a liquid in the ocean, a solid on land, and a vapor in the atmosphere. The substances can also be used by specific organisms at different times in the cycle. The important concept is that the substances are constantly moving through the systems—sometimes quickly (within hours, such as a rainstorm) or slowly (taking thousands of years, such as in a glacier).

THE CARBON CYCLE—NATURAL V. HUMAN AMPLIFICATION

Thus, the *carbon cycle* is extremely important. It also plays a critical role in global warming. Carbon dioxide enters the air during the carbon cycle. Because of its abundance, it enters from several sources. Vast amounts of carbon are stored in the Earth's soils, oceans, and sediments at the bottoms of oceans. Carbon is stored in the Earth's rocks and released when they erode. It exists in all living matter. Every time animals and plants breathe, they exhale CO_2.

When examining the Earth's natural carbon cycle, it is important to understand that the Earth maintains a natural carbon balance. Throughout geologic time, when concentrations of CO_2 have been disturbed, the system has always gradually returned to its natural (balanced) state. This natural readjustment works very slowly.

Through a process called diffusion, various gases that contain carbon move between the ocean's surface and the atmosphere. Because of this, plants in the ocean use CO_2 from the water for photosynthesis, which means that ocean plants store carbon, just as land plants do. When ocean animals eat these plants, they then store the carbon. Then when they die and decompose, they sink to the bottom, and their remains become incorporated in the sediments on the bottom of the ocean. Once in the ocean, the carbon can go through various processes. It can form rocks and shells. It can move around the ocean depths and exchange with the atmosphere.

As carbon moves through the system, different components can move at different speeds. Scientists break these reaction times down into two categories: short-term cycles and long-term cycles.

In short-term cycles, carbon is exchanged quickly. An example of this is evaporation, a gas exchange between the oceans and the atmosphere. Long-term cycles can take years or even millions of years to occur. Examples of this are carbon stored for years in trees or carbon weathered from a rock being carried to an ocean, being buried, incorporated into plate tectonic systems, then later being released into the atmosphere through a volcanic eruption.

Throughout geologic time, the Earth has been able to maintain a balanced carbon cycle. Now this natural balance has been upset by human

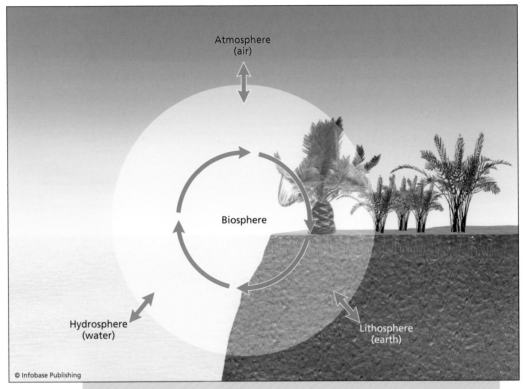

Human activity has an effect on the atmosphere, land, and water systems. Elements cycle in and out of each dynamic component.

activity. Over the past 200 years, fossil fuel emissions, land-use changes, deforestation, agricultural changes, etc., have increased atmospheric carbon dioxide by 30 percent (and methane, another greenhouse gas, by 150 percent) to concentrations not seen in the past 650,000 years.

Humans are adding CO_2 to the atmosphere much faster than the Earth's natural system can remove it. Prior to the Industrial Revolution, atmospheric carbon levels remained constant at around 280 ppm. This meant that the natural carbon sinks were balanced between what was being emitted and what was being stored. After the Industrial Revolution began and CO_2 levels began to increase—315 ppm in 1958 to 383 ppm in 2007—the balancing act became unbalanced and the natural sinks could no longer store all the carbon that was being introduced into

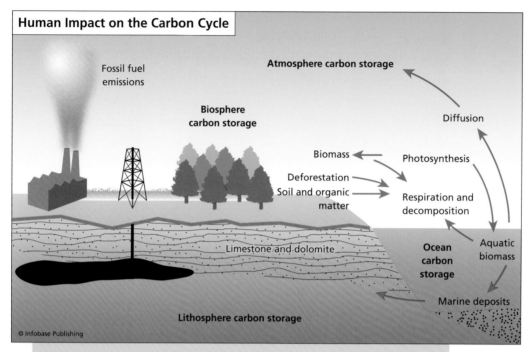

Human Impact on the Carbon Cycle

Fossil fuel emissions

Atmosphere carbon storage

Biosphere carbon storage

Diffusion

Biomass ← Photosynthesis

Deforestation

Soil and organic matter

Respiration and decomposition

Limestone and dolomite

Ocean carbon storage

Aquatic biomass

Marine deposits

Lithosphere carbon storage

© Infobase Publishing

Human beings affect the Earth's natural carbon balance by adding more carbon than it can handle.

the atmosphere by human activities. In addition, according to Dr. Pep Canadell of the National Academy of Sciences, 50 years ago for every ton of CO_2 emitted, 1,323 pounds (600 kg) were removed by natural sinks. In 2006, only 1,213 pounds (550 kg) were removed per ton, and the amount continues to fall. This indicates that the natural sinks are losing their carbon storage efficiency. This means that while the world's oceans and land plants are absorbing great amounts of carbon, they cannot keep up with what humans are adding. The natural processes work more slowly than the human-influenced ones do. The Earth's natural cycling usually takes millions of years to move large amounts from one system to another. The problem with human interference is that the introduction of large amounts of CO_2 is happening in only centuries or decades—and the Earth cannot keep up. The result is that each year the CO_2 concentration of the atmosphere gets higher, making the Earth's atmosphere warmer.

The problem began with the Industrial Revolution. This was the period in the late 18th century when industrialization began and changed lifestyles with the invention of steam power, electricity, mechanization, development of fossil fuels, etc.

Before the Industrial Revolution, the primary CO_2 produced came from natural processes, such as decaying biomass or burning wood, and humans and animals, who give off CO_2 when they breathe. And the Earth's natural carbon cycle was able to keep the carbon cycle in balance. The oceans and vegetation took it in and released oxygen back into the system.

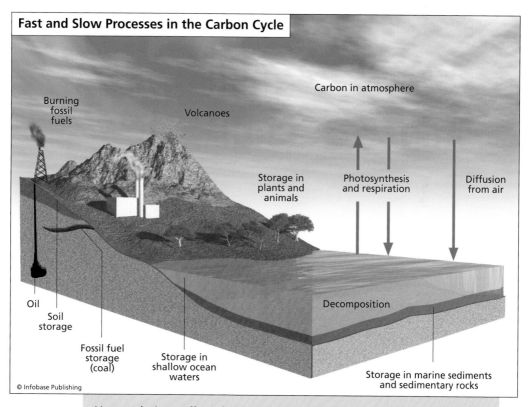

Fast and Slow Processes in the Carbon Cycle

Carbon in atmosphere

Burning fossil fuels

Volcanoes

Storage in plants and animals

Photosynthesis and respiration

Diffusion from air

Oil

Soil storage

Decomposition

Fossil fuel storage (coal)

Storage in shallow ocean waters

Storage in marine sediments and sedimentary rocks

© Infobase Publishing

Human beings affect the Earth's natural carbon cycle through the burning of fossil fuels and deforestation, among other things. The natural system cannot keep a balance with human emissions involved. *(modeled after IPCC)*

With industrialization, factories were built to produce thousands of products, electricity was generated to make all the systems operate, and fossil fuels were invented in order to transport goods. The past two centuries have put a huge stress on the environment. Besides the fossil fuels that we have been using in ever-increasing amounts, deforestation contributes to global warming. By burning and cutting down the forests, we are destroying their ability to store carbon and we are releasing stored carbon back into the atmosphere, exacerbating the situation.

THE HYDROLOGIC CYCLE

The hydrologic (water) cycle, like other cycles, plays a direct role in the function of healthy ecosystems. The hydrologic cycle describes the movement of all the water on Earth. It has no starting point and involves the existence and movement of water on, in, and above the Earth. The Earth's water is always moving and changing states—from liquid to vapor to ice and back again. This cycle has been in operation for billions of years, and all life depends on its existence. Many scientists are concerned because global warming affects major components of the water cycle in a negative way.

When the cycle is in equilibrium, water is stored as a liquid in oceans, lakes, rivers, in the soil, and underground in spaces in the rocks (called aquifers). Frozen water is stored in glaciers, ice caps, and snow. It is also stored in the atmosphere as water vapor, droplets, and ice crystals in clouds. As noted previously, water can change states and move to different locations. For instance, water can move from the ocean to the atmosphere when it evaporates and turns from a liquid to a gas (vapor). Plants release water as a gas through transpiration. Water in the atmosphere condenses to form clouds, which can then form rain, snow, or hail and return to the Earth's surface. Water that comes back to Earth can be stored where it lands (in an ocean) or it can flow aboveground (river), or it can infiltrate the soil and move underground (as groundwater).

This is how the Earth's natural water cycle works. Once global warming kicks in, however, it enhances the water cycle, making it more extreme. Global warming causes the Earth's atmosphere to warm, evap-

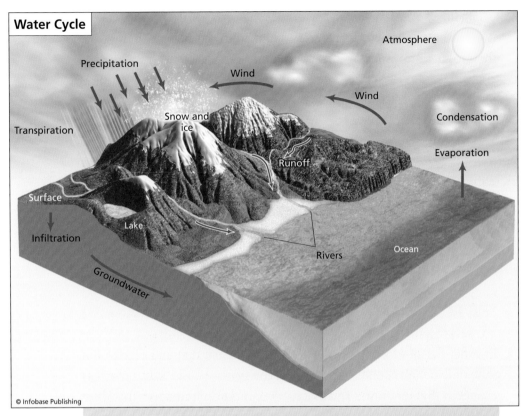

Water Cycle

Atmosphere

Precipitation

Wind

Wind

Condensation

Snow and ice

Transpiration

Evaporation

Runoff

Surface

Lake

Infiltration

Rivers

Ocean

Groundwater

© Infobase Publishing

Water continually cycles through the lithosphere, hydrosphere, biosphere, and atmosphere. Water evaporates into the atmosphere from the land and from the oceans and lakes. Plants and animals use and reuse water and release water vapor into the air. Water vapor in the air condenses to form clouds, which can turn to precipitation and fall back to the Earth.

orating more water and holding more water (the warmer the air, the more water it can hold).

This may have a twofold effect. First, in areas where more water vapor exists, more clouds will form, causing more rain and snow. In other areas, especially those far away from water sources, more evaporation and transpiration (together called evapotranspiration) could dry out the soil and vegetation. This means there would be fewer clouds and less precipitation, which could cause drought and cause huge problems for farmers, ranchers, cities, and wildlife habitat. All ecosystems in these areas would be negatively affected.

In areas receiving increasing amounts of precipitation, such as portions of Japan, Russia, China, and Indonesia, more water will infiltrate the ground and over the surface. It could increase the levels of lakes and rivers, causing serious flooding and even forming new lakes. Wetter conditions will also affect the plants and animals in the area.

Drier areas, such as Africa and the southwestern United States, will also experience serious effects. As the ground dries out from evapotranspiration, the atmosphere loses an important source of moisture. This, in turn, creates fewer clouds, which means there is less rain, making the area more arid. As less water is available to infiltrate the ground, less will be able to live off, on, or in the soil. Rivers and lakes would dry up, vegetation would die off, and the land would no longer be able to support humans, animals, and other life.

THE RELATIONSHIPS BETWEEN THE LAND, THE OCEANS, AND THE ATMOSPHERE

In order for all of Earth's cycles to work together, they must exist in a state of dynamic equilibrium. This means that as substances move and change at different times and places, they must do it in a way that does not have a negative effect on the entire working system—all the components must work together and complement each other.

This dynamic equilibrium changes as the seasons change, because different needs must be met at different times. In the spring and summer when plants grow, they need carbon so they take it from the atmosphere and the soil. Then, when the growing season is over in the fall and winter, plants release carbon back to the soil and atmosphere. This is significant especially in the Northern Hemisphere because most of the landmasses are located there, creating a global seasonal change of CO_2 in the atmosphere.

The oceans and atmosphere also interact extensively. Oceans are more than a moisture source for the atmosphere. They also act as a heat source and a heat sink (storage), as well as a carbon sink.

CARBON SINKS AND SEQUESTRATION

A carbon sink is a reservoir that accumulates and stores carbon. Natural carbon sinks include the oceans and the photosynthesis of plants.

The process whereby these sinks remove carbon from the atmosphere is known as carbon sequestration. This is an important concept for global warming because it is becoming increasingly urgent that we find ways to store the extra CO_2 that is being added to the environment because of human activities.

Forests are carbons sinks. Trees take in CO_2 as part of the photosynthesis cycle and then store the carbon in the plant and soil. In fact, scientists have determined that about half of a tree's weight is carbon. When trees live for many years, this carbon is effectively stored for long periods of time. It is estimated that vegetation stores 600 billion tons of CO_2. Based on research at NASA using high-resolution maps of carbon storage, they determined that the forests in the United States, Europe, and Russia stored almost 700 million metric tons of carbon each year during the 1980s and 1990s.

The U.S. Environmental Protection Agency (EPA) is currently focusing attention on carbon sequestration efforts through agriculture and forestry in order to help prevent global climate change. They are focusing on maintaining existing forests in order to keep the carbon already there safely stored. According to the EPA, "Forests and soils have a large influence on atmospheric levels of CO_2. Agricultural and forestry activities can both contribute to the accumulation of greenhouse gases in the atmosphere, as well as be used to help prevent climate change by avoiding further emissions and by sequestering additional carbon." A second approach is to increase the carbon storage capacity in the United States by planting more trees and by modifying some agricultural practices, such as having farmers convert from conventional tillage to conservation tillage methods. They also support the substitution of bio-based fuels and products for fossil fuels (coal and oil) because these alternative fuels produce less CO_2 when used.

The oceans also function as a carbon sink. Carbon dioxide dissolves in the ocean from the air, and marine animals also extract CO_2 to support their own needs. Depending on the temperature and pressure of the water, the ocean is able to absorb and dissolve large amounts of CO_2. Scientists have estimated that 1,020 trillion tons (925 trillion metric tons) of CO_2 is stored by the surface waters, and 40 billion tons (36 billion metric tons) in the deep waters.

Carbon dioxide is also stored in the Earth's soil and rocks, although not as much is stored here as in the oceans and vegetation. Methods of sequestration are being looked at more seriously as one way to counter the effects of global warming.

The U.S. National Energy Technology Laboratory's (NETL) Carbon Sequestration Program overseen by the U.S. Department of Energy (DOE) is helping to develop technologies to capture, purify, and store CO_2 in order to reduce greenhouse gas emissions without having an adverse impact on energy availability and use or harming economic growth. NETL sees this as an important issue because they have determined that worldwide CO_2 emissions generated from human activity have increased from insignificant levels 200 years ago to more than 33 billion tons (30 billion metric tons) today. In fact, the U.S. Energy Information Administration (EIA) predicts that if no action is taken, the United States will emit 7,550 million tons (6,850 million metric tons) of CO_2 by 2030, which means an increase in emission levels from 2005 of more than 14 percent.

It is NETL's goal to have in place by 2012 technologies that provide safe, cost-effective, commercial-scale greenhouse gas capture, storage, and mitigation. Technologies being developed include injecting CO_2 into geological formations, increasing carbon uptake on mined lands, using no-till agriculture, reforestation, rangeland improvement, wetlands recovery, and riparian restoration. The NETL is also involved in research into high-speed computing, simulations, and modeling as tools for designing, optimizing, analyzing, and better understanding the chemical and physical processes that take place in carbon sequestration.

An international climate change initiative has been established, the Carbon Sequestration Leadership Forum (CSLF), composed of 21 countries and the European Community. Its goal is to develop improved, cost-effective technologies for CO_2 capture and long-term storage. Justin R. Swift, deputy assistant secretary for international affairs for fossil energy at the DOE, says that "CSLF members also work to make the technologies broadly available internationally, to help developing countries learn about and apply the technologies, and to identify and address regulatory and policy issues that relate to carbon capture and storage." According to the CSLF, "Joint projects are already increasing knowledge in areas that include technology, economics, health, safety, and the environment, and also demonstrate a wide range of CO_2 capture, transport, and storage research and activities."

Swift says, "The CSLF has recognized ten projects worldwide. Two of these are: (1) the Weyburn CO_2 Monitoring and Storage Project, a collaboration among the United States, Canada, the European Commission, and Japan; and (2) the Frio Brine Pilot Experiment, a joint project between the United States and Australia." In the Weyburn Project, the Weyburn oil field in southern Saskatchewan uses a CO_2-enhanced oil recovery method to extract the oil. They acquire CO_2 from the Great Plains Gasification Plant in North Dakota and pump it into the ground at Weyburn. Injecting the CO_2 serves two purposes: The injection of CO_2 reduces the oil's viscosity and expands its volume, allowing more oil to be withdrawn, the injected CO_2 stays safely underground.

According to S. Julio Friedmann, who heads the Carbon Storage Initiative at the DOE Lawrence Livermore National Laboratory in California, "Weyburn combines carbon sequestration and enhanced oil recovery." This project also is allowing the development of a cutting-edge monitoring and tracking technique leading to a better understanding of CO_2 movement in the new CO_2 storage reservoir.

About the Frio Brine Pilot, Judd Swift says, "Its purpose is to ensure safe storage in a saline reservoir. We consider it a key U.S. project for CO_2 injection. Saline reservoirs show particular promise worldwide for carbon sequestration." Frio Brine researchers drilled a well in 2004 and injected 1,904 tons (1,727 metric tons) of CO_2 4,921 feet (1,500 m) underground at the South Liberty oil field near Dayton, Texas. The goal of this project is monitoring and research to gain an understanding of CO_2 storage for the future in saline aquifers.

According to Julio Friedmann, "The Frio formation is an enormous aquifer on the Texas gulf coast. They wanted to demonstrate that they could store CO_2 in that aquifer and, based on what they learned, demonstrate the potential for a large number of carbon storage projects along the Texas gulf coast."

Large-scale injection of CO_2 for the sole purpose of removing it from the atmosphere is thus far only taking place in Norway, at the offshore Sleipner facility, which is run by Statoil, the oil company owned and run by Norway. Statoil strips excess CO_2 from gas and injects it into an aquifer 2,625 feet (800 m) below the seabed. Operational since 1996, they have injected at least 1 million tons of CO_2 a year at a low cost. This offers hope for the future.

Plate Tectonics: Climate and Movement of the Earth's Continents

The continents of the Earth have not always been in the same geographical positions that they are today. This chapter discusses the theory of plate tectonics, also known as continental drift, as well as scientists' theories of how specific movements of the continents have affected the Earth's climate and global warming throughout history.

THE THEORY OF CONTINENTAL DRIFT

At one point in geologic time, the world was made up of a single continent called Pangaea. Over time, this supercontinent separated and drifted apart, forming the different continents that exist today. The process is always in motion; the plates always moving, as they will continue to do far into the geologic future. In fact, the configuration of the Earth's landmasses may change drastically from the way they are today.

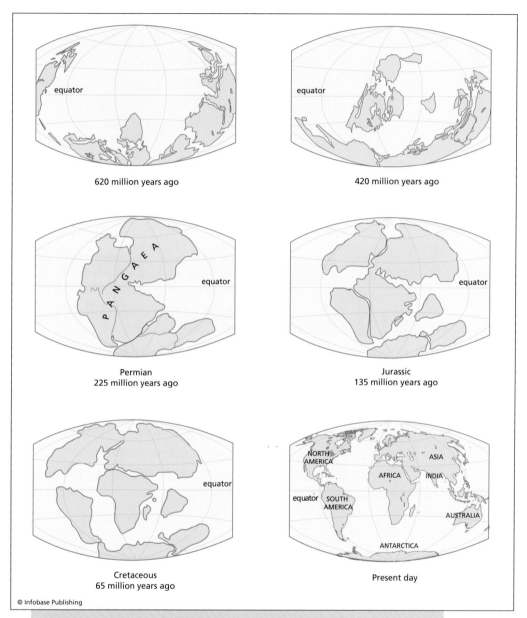

The Earth's plates have shifted over geologic time.

In 1912, Alfred Wegener, a German scientist, geologist, and meteo-rologist, was the first to propose the theory that the Earth's landmasses were not stable, but instead drifted slowly across the Earth's surface. He used the theory of Pangaea's existence along with field data collected from rocks. He believed that the continents are made of lighter rocks that rest on heavier crustal material—similar to the way an iceberg floats in water. Wegener proposed that the positions of the continents are not rigidly fixed, but instead move at a slow rate—about one yard per century. It was not until the 1960s that geologists gained the technology to understand the processes at work that move the Earth's plates. The concepts of seafloor spreading and plate tectonics emerged as powerful new hypotheses to explain the movements of the Earth's surface. Scientists concluded that the Earth's surface was not composed of one large sheet, but rather of more than 12 major pieces of crust called plates that fit together like pieces of a jigsaw puzzle.

Each rigid plate, or slab, of lithosphere averages at least 50 miles (80 km) thick. They move relative to one another at speeds of a few inches per year—about the same rate as a human fingernail grows. Although these rates are slow by human standards, they are very rapid by geologic ones, where processes can take millions or billions of years. Plates can move 30 miles (50 km) in 1 million years, and they have already been in motion for 100 million years. Scientists recognize three common boundaries between the moving plates: divergent, convergent, and transform.

Divergent

These plates move away from each other. Their boundary is characterized by a chasm, filled with molten rock from within the Earth. The best-known divergent boundary is the Mid-Atlantic Ridge, a 12,000-mile (19,300-km) long part of a submerged mountain range that extends for 42,000 miles (67,600 km). This underwater mountain chain is four times longer than the Andes, the Rockies, and the Himalayas combined.

Convergent

These plates move toward each other and collide. One of the plates is dragged down, or subducted, beneath the other. The area where the plate sinks under the adjacent plate is called the subduction zone. Convergence can occur between oceanic-continental plates, oceanic-oceanic plates, and continental-continental plates. An example of convergent plates is the Nazca plate (oceanic) subducting under the South American plate and creating the majestic Andes Mountains in South

America. When plates push together, the Earth's crust tends to buckle and be pushed upward or sideways. This is how the Himalayas was formed. Towering as high as 29,000 feet (8,854 m), they form the highest continental mountains in the world. Strong, destructive earthquakes and rapid uplift of mountain ranges are common. When an oceanic plate subducts under another oceanic plate, a trench is formed. Trenches can be hundreds of miles long and five to seven miles (8 to 10 km) deep cutting into the ocean floor.

Transform

These plates slide horizontally. Most transform faults are found on the ocean floor, offsetting some of the active spreading ridges, producing zigzag plate margins. Shallow earthquakes are associated with them. The best-known example of a sliding plate is the San Andreas Fault in California. The San Andreas, which is 800 miles (1,300 km) long and in places up to 50–100 miles (80–160 km) wide, slices through two-thirds of the length of California. Here, the Pacific plate has been grinding horizontally past the North American plate for 10 million years, at an average rate of two inches (5 cm) per year. Land on the west side of the fault zone (Pacific plate) is moving in a northwesterly direction relative to the land on the east side of the fault zone (North American plate). Because the plates are internally rigid, moving as a solid mass, they interact mostly at their edges. All plates move relative to each other, by sliding past or subducting under each other.

THE FORCE THAT DRIVES THE TECTONIC PLATES

Tectonic plates do not randomly drift or wander about the Earth's surface; definite, yet unseen, forces drive them. Scientists believe that the relatively shallow forces driving lithospheric plates are also working with forces that originate much deeper in the Earth.

From seismic and other geographical evidence and laboratory experiments, scientists generally agree with Harry Hess's theory that the plate-driving force is the slow movement of the hot, softened mantle that lies below the rigid plates. Scientists also accept that the circular motion of the mantle carries the continents along, much like a conveyor belt. As John Tuzo Wilson stated in 1968, "The Earth, instead of appearing as an inert statue, is a living mobile thing." Both the Earth's surface and interior are in motion.

Below the lithospheric plates, the mantle is partially molten and can slowly flow in response to steady forces applied for long periods of time. When solid rock in the Earth's mantle is subjected to heat and pressure in the Earth's interior over millions of years, it can be softened and molded to different shapes.

Within the mantle, the movement is as a convection cell in a circular motion, similar to heating a pot of thick soup to boiling. The heated molten mantle material rises toward the Earth's crust, spreads, and begins to cool, and then sinks back toward the Earth's core, where it is reheated, rises again, and repeats the process. Convection in the Earth is very slow. In order for convection to occur, there must be a source of heat. Heat within the Earth comes from two main sources: radioactive decay and residual heat.

Radioactive decay is a spontaneous process that involves the loss of particles from the nucleus of an isotope (the parent) to form an isotope of a new element (the daughter). An isotope is one variation of an element, different from other variations by its unique number of neutrons. Radioactive decay of naturally occurring chemical elements (such as uranium, potassium, and thorium) releases energy in the form of heat, which slowly migrates toward the Earth's surface. Residual heat is gravitational energy left over from the formation of the Earth 4.6 billion years ago.

How and why the escape of interior heat becomes concentrated in certain regions to form convection cells still remains largely a mystery. Scientists do believe that plate subduction plays a more important role than seafloor spreading in shaping the Earth's surface features and causing the plates to move. The gravity-controlled sinking of a cold, denser oceanic slab into the subduction zone, dragging the rest of the plate along with it, is considered to be the driving force of plate tectonics.

Forces deep within the Earth's interior drive plate motion. Because these powerful forces are buried so deeply, no mechanism can be tested directly and proven beyond a doubt. The fact that the tectonic plates have moved in the past and are still moving today is certain, but the details of why and how they move will continue to challenge scientists in the future.

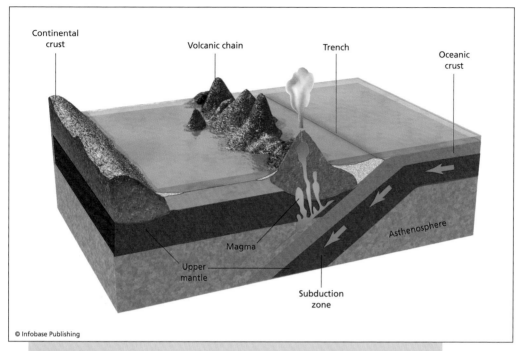

Continental crust

Volcanic chain

Trench

Oceanic crust

Magma

Asthenosphere

Upper mantle

Subduction zone

© Infobase Publishing

As the plates move, they create characteristic landforms on the Earth's surface.

Because the Earth's plates have been in motion for millions of years, they have moved hundreds of miles. Seafloor spreading over the past 100 to 200 million years has caused the Atlantic Ocean to grow from a tiny inlet of water between the continents of Europe, Africa, and the Americas into the vast ocean of today.

Oceanic trenches are the deepest parts of the ocean floor. One of the most famous trenches is the Mariana Trench. The faster-moving Pacific plate converges against the slower-moving Philippine plate. The Challenger Deep, at the southern end of the Mariana Trench, is the deepest part of the ocean and plunges deeper into the Earth's interior—36,000 feet (11,000 m)—than Mount Everest, the world's tallest mountain, rises aboveground.

Oceanic-oceanic plate convergence also results in the formation of volcanoes. Over millions of years, the erupted lava builds up on the ocean floor until the submarine volcano rises above sea level to become an island volcano. Earthquakes are common in these areas as well. The Pacific Ring of Fire, a string of active volcanoes around the Pacific basin, is the world's most important example of oceanic-oceanic convergence.

Not all plate boundaries are as simple as the three types mentioned above. In some regions, the boundaries are not well defined because the plate movement deformation extends over a broad belt—called a plate boundary zone. These areas typically have larger plates with several smaller fragments of plates, called microplates, involved. One example of this is the Mediterranean-Alpine boundary, which involves two major plates and several microplates. Another microplate is the Juan Fernandez at the Pacific-Nazca-Antarctic junction.

PLATE TECTONICS AND GLOBAL WARMING

Plate tectonics is one of the three major types of climate forcing in the natural world (the other two being changes in the Earth's orbit and changes in the strength of the Sun). Climate forcing is when a mechanism "forces" the climate to change. These are both natural and human-caused climate forcing mechanisms. Unlike some forces that have a direct effect on climate, such as the atmosphere, the land's surface, and vegetation, this one works very slowly, over millions of years.

Because scientists have been able to reconstruct the past positions of continents and the shapes of ocean basins, it has allowed them to identify periods of global cooling (icehouse) intervals, when massive ice sheets covered much of the Earth, and greenhouse intervals, when no ice existed. In addition, when scientists can accurately measure plate tectonic processes as they have changed the Earth's surface and then compare these changes to the climate that occurred at that point in geologic time, they can determine possible cause-and-effect relationships between changes in the Earth's tectonic system and its climate.

One of the most helpful characteristics that has allowed scientists to look back through the window of time and reconstruct the ancient positions of the continents is the Earth's natural magnetic field. Some of the Earth's molten rocks, such as basalts, are rich in highly magnetic iron. When they cool, the iron-rich elements orient themselves to the magnetic north, turning them into fossil compasses. When scientists use the magnetic orientations of the rocks to determine how the rock was positioned relative to north at the time the formation was created, they can recon-

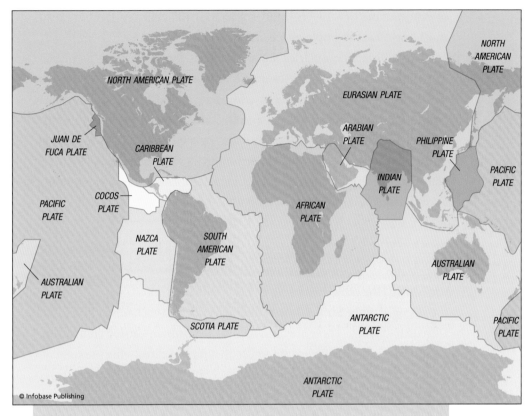

These are the major tectonic plates of the world.

struct the past positions of continents and ocean basins. This discipline is called *paleomagnetism*.

Paleomagnetism has been successfully used to reconstruct movements of plates and rates of seafloor spreading. The drawback to this technique is that it can only date rocks back 175 million years, because that is the age of the oldest rock; anything older has already been subducted beneath a continental plate and remelted in the plate tectonic process. In order to go back farther in time, scientists must look at rock records on continents.

The technique is also limited to continental shifts in latitude, not longitude. This is because when continents are at the equator, their compass setting will be horizontal to north and at the North Pole the compass will be nearly vertical. This is because the equator is the farthest point away from the poles, giving the largest declination. Longitudinally, it does not matter because the longitude reading does not affect the north/south reading on the compass (because it is a constant north/south direction).

Throughout the Earth's history, the Earth's magnetic field has repeatedly reversed direction. For example, compasses that point to magnetic north today would have pointed to magnetic south (a position near the Earth's South Pole) during those intervals when the field was in a reversed orientation. These reversals are irregular. Sometimes they do not happen for several million years, other times they may happen after a few thousand years.

When scientists drill core samples out of basaltic rock and find evidence of these magnetic reversals, they can use the information to determine past continent location and climate change. Other methods of determining past locations of the Earth's continents come from analysis of landforms, similar to fitting the pieces of a jigsaw puzzle together. Overall, scientists can reconstruct the positions of continents on the Earth's surface with good accuracy back to about 300 million years ago, and less accurately back to 500 million years ago.

Concerning plate tectonics, climate, and global warming, three theories have been proposed: (1) the polar position hypothesis, (2) control of CO_2 by seafloor spreading, and (3) control of CO_2 by uplift and weathering.

The Polar Position Hypothesis

This idea was suggested early on and said that ice sheets should appear on continents when they are located at polar, or near-polar, locations (high latitudes). If there were not any continents near the North or South Poles, then no ice should exist anywhere on Earth. This theory only takes into effect movement of the continents through plate tectonics.

While this theory has proven true for many demonstrated time intervals in the past, it has not always been true. Today is one example. There are ice age sheets at high latitudes (polar ice caps and ice sheets), that exist because of (1) cold temperatures caused by low angles of sunshine, (2) high albedoes (reflectivity) because of the snow and sea ice cover, and (3) adequate snow being added on a regular basis to the ice sheets. But there are also glaciers that exist in many other places in the world that are not at high latitudes.

During a time interval from 425 to 325 million years ago, land existed at the South Pole for almost 100 million years without any ice sheets forming. Although this is one hypothesis, it illustrated that there was more involved in determining cooling and warming periods than just the polar position of continents. It introduced the concept that there had to be other inputs necessary to explain large-scale glaciations. By experimenting with general circulation models (GCMs), it was proposed that periods of warmer climate were caused by elevated levels of CO_2 in the atmosphere.

Several paleoclimatic studies concerning Pangaea support this theory. It has been determined that no ice sheets existed on Pangaea 200 million years ago, even though the supercontinent's northern and southern limits lay well within the Arctic and Antarctic Circles. These areas correspond with areas today like Greenland, which is ice covered. Because Pangaea did not have any polar ice, paleoclimatologists believe that Pangaea's climate was warmer than the Earth's climate is today.

Fossil evidence of trees dating back to the time of Pangaea also supports this theory. A variety of different palmlike vegetation grew at latitudes as high as 40 degrees. This could not happen today—winters would freeze them. Using evidence like this indicates that the Earth had a much more tropical environment at higher latitudes during this time period. This leads to a conclusion that the reason for a warmer Earth during this time period is that the CO_2 level was much higher then than it is today. Many different climate models have been run using different thresholds of CO_2 levels in the atmosphere and compared to the results of the geologic evidence from Pangaea. It has been determined that during this period the atmospheric level of CO_2 was 1,650 ppm, nearly six times the level of CO_2 before the Industrial Revolution (280 ppm).

Climate models also predict high aridity in the interior of Pangaea due to the high CO_2 content. Geologic evidence also supports the model's predictions of widespread aridity. A major source of evidence is the existence of an extensive distribution of evaporite deposits. Evaporite deposits are salts that precipitate out of water that does not drain into the ocean. This occurs when evaporation is greater than precipitation. Because there is no outlet to an ocean, when the water evaporates, it leaves the salts behind as a deposit. A well-known example of this today is the Bonneville Salt Flats in Utah. It is a remnant of ancient Lake Bonneville, a large lake that existed from about 32,000 to 14,000 years ago. It occupied the lowest, closed depression in the eastern Great Basin and at its largest extent covered about 20,000 square miles (51,800 km^2) of western Utah and smaller portions of eastern Nevada and southern Idaho. The lake eventually breached and drained during the last ice age. Today, the only remnants of this massive lake are the Great Salt Lake and the Bonneville Salt Flats.

The Salt Flats are massive evaporite deposits that extend for miles. They are extremely flat—one of the flattest places in the world. Geologists have determined that more evaporite salt was deposited at the time of Pangaea than at any time in the last several hundred million years. Because most of these deposits exist in what would have been the interior of the Pangaen continent, this gives climatologists valuable insight into what will happen to the interior areas of large continents if CO_2 levels increase and global warming intensifies.

Paleoclimatologists also believe that the warmer climate of Pangaea kept the temperatures so far above freezing that any snow melted quickly, never accumulating long enough to allow glaciers to form.

Control of CO_2 by Seafloor Spreading

The seafloor spreading theory is one hypothesis introduced to suggest a mechanism of controlled CO_2 levels in the atmosphere, causing the observed variations between warm CO_2-rich greenhouse intervals and cold CO_2-depleted global cooling intervals. This theory focuses the control of CO_2 levels on plate tectonic processes. During plate tectonics, carbon is cycled endlessly between the Earth's interior and its surface. It is this cycling of carbon in different stages that defines whether global warming or global cooling intervals predominate.

Rock deep within the Earth exists as magma, which contains a rich supply of gases. As the magma rises to the surface during volcanic activity, CO_2 escapes at the plate margins—such as at oceanic ridges where volcanoes are commonly found. Most CO_2 enters the atmosphere at the margins of divergent plates where the hot magma carrying CO_2 erupts right into the ocean water. These areas are the *ocean ridges*. One of the best known is the Mid-Atlantic Ridge. The other likely place CO_2 can enter the atmosphere is at the margins of converging plates, where the subducting plates melt and form molten magma that rises to the surface in mountain belt

This is the Bonneville Salt Flats in Utah. It is a large, flat region of pre-cipitated salt; a remnant of an ancient inland sea. Scientists believe this is what the interior of Pangaea looked like. *(USGS)*

and island arc volcanoes. These active volcanoes actively spew CO_2 into the atmosphere around them. Volcanoes formed by hot spots in the ocean floor are another location where large amounts of CO_2 can be released into the atmosphere. The Hawaiian Islands are an example of this. Formed over a stationary hot spot, as the oceanic plate travels northwestward, the island chain is formed one island at a time.

Scientists have also determined that the rate of CO_2 added to the atmosphere by these processes is also controlled directly by the speed at which continental drift takes place. The rate of spreading determines how much CO_2 is released to the atmosphere. When seafloor spreading speeds up, more lava is produced, causing more CO_2 to be released into the atmosphere, promoting global warming. Faster spreading rates at the mid-ocean trenches likewise cause the subduction zones (an ocean plate being pushed under a continental plate and melting) to speed up, which delivers larger amounts of carbon-loaded sediments and rock to be sub-ducted and melted again inside the Earth, where volcanoes then form, releasing significant amounts of CO_2 into the atmosphere to cause global warming. When plate movements are slower, less volcanic activity occurs, so less CO_2 is added to the air, preventing overheating. These conditions would correspond to the global cooling intervals.

The rate of seafloor spreading changes over tectonic intervals of tens of millions of years, affecting the rate of CO_2 transfer from the Earth's rock reservoirs to its atmosphere. One example occurred approximately 100 million years ago, when the average spreading rate was much faster than it is today. Because of this, the rate of input of CO_2 to the atmosphere was much higher, creating a greenhouse world. This theory also matches existing geological data. Ice sheets, for example, did not exist at that time, and this theory offers a possible explanation for that.

Control of CO_2 by Uplift and Weathering

This theory looks at natural uplift and *weathering* of the Earth's surface features as the determining factor of CO_2 levels in the atmosphere and therefore potential global warming. When land is uplifted, geologic forces go to work on it to erode and wear the land down—a process called weathering.

The Earth's surface can be uplifted in several different ways. Earthquakes can fracture, fault, and uplift large sections of the Earth's crust. Plate tectonics can also create mountains and high plateaus when converging plates push against each other. Once land has been uplifted, it is a natural geomorphic process to have various erosional processes begin working on the landscape to weather the rock, erode formations, transport sediments, and wear the Earth's surface down to a flat plain. Geomorphology is the study of the nature, origin, and development of landforms and the changes they go through during mechanical, physical, and chemical processes. This is a continual cycle because when a flat area is uplifted again by geologic forces, geomorphic processes go to work again on the landscape under the forces of gravity to erode it to a flat level again.

Several processes can weather and erode landscapes. Steep slopes are subject to *mass wasting*, the downslope movement of material under the force of gravity. Mass wasting processes include avalanches, mudslides, rockslides, rock falls, debris flows, and mudflows. Depending on the process, material of all sizes can be dislodged and moved from huge slabs of rock to boulders, rocks, gravel, and soil. When mass wasting occurs, fresh rock is exposed, which is then vulnerable to chemical weathering pro-

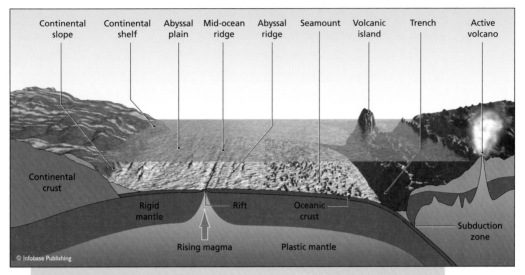

Plate tectonics can add CO_2 to the atmosphere through volcanic sources and seafloor spreading. Weathering of rock also adds CO_2. Scientists believed these mechanisms played a role in the CO_2 budget by increasing CO_2 levels in the atmosphere.

cesses from rain and snowstorms. The more precipitation a slope receives, the more chemical weathering occurs. Glaciers can also erode rocks and leave them vulnerable to weathering.

Scientists have suggested that the uplift weathering hypothesis is important because faster rates of weathering take more CO_2 out of the atmosphere, which then cools global climate, leading toward global cooling conditions. Scientists have successfully applied this hypothesis to uplift of the Earth's surface created by the collision of continents due to plate tectonics.

Interestingly, both the seafloor spreading and uplift-weathering hypotheses fit well with other evidence, such as the presence of ice sheets, orogenic uplift, and corresponding chemical weathering, that has been found dealing with identified global cooling/global warming intervals over the past 400 million years. When scientists are able to make these connections with processes that continue today, it allows greater insight and better understanding of the complexity of processes involved in global warming. This, in turn, allows scientists to improve climate models, enabling better management and resource decisions to be made.

The Flow of Energy

Global warming is tied directly to the flow of energy from various sources to, in, and around the Earth. It involves energy from the Sun in the form of electromagnetic radiation, energy transfer and emission from the Earth, and the variations in flow of energy in the atmosphere to drive the Earth's energy balance. This chapter examines the flow and interaction of various sources of energy and its connection to global warming.

SOLAR ENERGY

The Sun is the source of most of the energy that drives the climate systems on Earth. The nuclear reaction processes occurring within the Sun have caused it to expand over time and gradually become brighter. Models indicate that the earliest Sun shone 25 to 30 percent more faintly than today; its luminosity has slowly increased to what it is today. This scenario has created an intriguing problem for climate scientists, however. A decrease of just a few percentage points of luminosity of the present-day Sun's current strength would cause all the water on Earth to freeze, despite the warming of the natural greenhouse effect. If all the water froze, the resultant high albedo would keep them from melting. Climatic models simulating this weak Sun with greenhouse gases on Earth at their present-day values show that Earth would have been frozen for the first 3 billion years of its existence.

The geologic record of ancient Earth's climate, however, does not show this to be the case. Evidence exists of geologic formations caused by running water, which indicates an unfrozen planet. The first evidence of deposition from glaciers in polar regions does not show up in the geologic record until 2.3 billion years ago. This evidence of an unfrozen Earth is also supported by the continued presence of life on Earth, which is found in the fossil record.

This period is referred to as the Faint Young Sun Paradox. It is still viewed today as a scientific mystery: with so weak a Sun, why was Earth not frozen for the first two-thirds of its history? The mystery is compounded with evidence of Snowball Earth (an Earth completely frozen) from 850 to 550 million years ago, when the Sun was only 5 percent weaker than today (unlike the early Sun, which was 30 percent weaker). In order to explain this, some scientists have suggested that greenhouse gases have acted like a thermostat, regulating the Earth's atmospheric temperature over time.

Today, the Sun produces incredible amounts of energy. It would take 440 million electric power plants to equal the energy coming to the Earth from the Sun. The intensity of the Sun, however, is not constant. Changes that occur on the inside of the Sun can affect the intensity of the sunlight that reaches the Earth's surface. If the intensity of sunlight increases, it causes warming on the Earth; conversely, if solar intensity weakens, it causes cooling on Earth. Scientists at NASA determined that the Sun's intensity weakened from 1500 to 1850. During this time, temperatures on Earth decreased about 2°F (1.2°C). The effect was especially noted in North America and Europe and is known by historians as the Little Ice Age.

The radiation coming from the Sun is short-wave radiation (as opposed to the energy given off by the Earth, which is considered long-wave radiation). There are cycles associated with energy output by the Sun. The sunspot cycles are some of the better-understood ones. They occur on roughly an 11-year cycle. Sunspots appear as dark areas and can appear in clusters of only a few to dozens. They last about a week, are caused by magnetic activity, and appear dark because they are cooler than the surrounding areas.

When sunspot activity is high, solar flares occur, which are bursts of high-energy particles. When these occur, the Earth gets bombarded with higher levels of solar radiation. It has been proposed that there is a connection to warmer or cooler climates on Earth being tied in some way to levels of solar activity; that perhaps levels of solar activity act as a trigger for specific climatic trends and conditions. The period from 1645 to 1715, for example, corresponds to the coldest portion of the Little Ice Age, called the *Maunder minimum*. It has been determined that this time period also had little or no sunspot activity.

When the Sun's radiation energy reaches the Earth's atmosphere, many things can happen to it. It can be reflected, absorbed, or transmitted. The Sun's energy arrives as electromagnetic radiation in the entire spectrum. When it reaches the atmosphere, the ozone layer absorbs the ultraviolet wavelengths (UV).

Ozone is a gas that occurs naturally in the Earth's atmosphere, specifically the layer called the stratosphere, several miles above the Earth's surface. It may exist in only very small amounts, but it serves a critical purpose to life on Earth. It acts as a shield against the Sun's harmful UV radiation that can cause skin cancer. Scientists have discovered that the use of chemicals, such as chlorofluorocarbons (CFCs), are harmful to natural ozone and actually depleted it, allowing an increase in harmful UV radiation to reach Earth. CFCs are organic compounds that contain carbon, chlorine, and fluorine atoms. They are very effective refrigerants that were developed to use in refrigeration units and air conditioners. The most common CFC was marketed under the trade name Freon. Because of their destructive effect on the ozone layer, their use has been banned in the United States.

When the cause of the depletion was discovered, international agreements were put into place to regulate their emissions. Because there has been international cooperation, scientists believe the ozone layer will eventually recover. It is this type of international cooperation that scientists would like to see happen with the global warming issue.

Water vapor and CO_2 absorb the infrared portion of the spectrum. Particles in the atmosphere scatter much of the Sun's incoming radiation. Of all the incoming energy, usually only 60 percent ever reaches the Earth's surface. The nature of the Earth's surface then determines what happens to the energy. Some of the energy is directly reflected back into space. The amount reflected by the Earth is determined by the

OZONE AND CLIMATE CHANGE

Scientists at NASA have determined that ozone and climate affect each other. According to Bill Stockwell of NASA's Desert Research Institute, temperature, humidity, winds, and the presence of other chemicals in the atmosphere influence ozone formation and the presence of ozone, in turn, affects those atmospheric constituents. When humans first began to harm the ozone layer in the early 1970s through the use of CFCs and halons (halons are a compound of one or two carbon atoms combined with bromine and one or more other halogens. Halons are gases used as fire-extinguishing agents. They are between three and 10 times more destructive to the ozone layer than CFCs.), causing a hole to form in the layer, research and international cooperation helped turn the harmful trend around. Scientists then predicted that under strict control it would be possible to see ozone levels completely recovered by the year 2050.

Recently, some startling new findings by NASA were introduced. They believe that even before the ozone's projected 2050 recovery, that ozone's effects on climate may become the main reason for ozone loss in the stratosphere. During the winter months, a vortex of winds develops around the polar regions, isolating the polar stratosphere. When temperatures drop below -109°F (-78°C), thin clouds of ice, nitric acid, surface's albedo. Ice caps and light color rocks, for example, have a high albedo, whereas dark soil and deep lakes have a low albedo. The texture of the surface and the angle of the Sun also determine how much energy is reflected or absorbed. A smooth texture or low Sun angle will reflect more energy.

In the last 35 years NASA scientists have determined that the amount of radiation the Sun emits during times of quiet sunspot activity has increased by .05 percent per decade. They believe this is important because if it increased steadily for many decades, it could cause significant climate change. Increased solar radiation leads to increased warming, which could add significantly to global warming. NASA's sci-

and sulphuric acid mixtures form. Chemical reactions on the surfaces of ice crystals in the clouds release active forms of chlorofluorocarbons (CFCs). Ozone depletion occurs, causing an ozone hole to appear. In the spring when temperatures begin to rise, the ice evaporates and the ozone layer begins to recover once again. Because of this, scientists are now warning that the ozone layer may not be fully recovered until 2060 or 2070.

Ozone influences climate through temperature. The more ozone that is in a given region of air, the more heat it retains. Ozone generates heat in the stratosphere in two ways: (1) it absorbs the Sun's UV radiation (which is especially helpful to humans for protection against harmful sunrays), and (2) it absorbs the longer infrared radiation that is reflected up from the lower atmosphere (the troposphere).

Decreased ozone in the stratosphere lowers temperatures. It has been determined that the mid to upper stratosphere has cooled by 2 to 11°F (1–6°C). This cooling period corresponds to when greenhouse gas amounts in the lower atmosphere (troposphere) have risen. Scientists at NASA think these two phenomena may be connected. The exact link between climate, ozone, and temperature is one that climatologists want to build more powerful models for.

Source: NASA, Goddard Institute for Space Studies

entists said that if this trend continued for a century or more it could significantly affect the outcome of global warming. Total solar irradiance (TSI) is the radiant energy received by the Earth from the Sun, over all wavelengths, outside the atmosphere. TSI interaction with the Earth's atmosphere, oceans, and landmasses is the biggest factor determining our climate. Even relatively small changes in this could have significant climatic effects.

THE EARTH'S ENERGY

Chapter 3 illustrated how the Earth's energy drives plate tectonics, affecting CO_2 levels and interactions with air masses in the atmosphere. Related to this, the energy of volcanoes can also affect climate because volcanoes emit both aerosols and CO_2 into the atmosphere. When a volcano erupts, it can send ash and sulfate gases to great heights in the atmosphere. If the sulfate combines with water, it produces tiny droplets of sulfuric acid, called aerosols. Aerosols are very small solid particles or liquid droplets dispersed in a gas—usually air.

When aerosols enter the atmosphere, they tend to block the Sun's incoming energy. If sunlight cannot reach the Earth's surface, this causes a cooling effect. Fortunately, aerosols do not stay in the atmosphere for extremely long periods of time like certain greenhouse gases do. Some particulates such as ash are big enough that they settle out of the atmosphere quickly. The finer the particulates, the longer their potential time in the atmosphere. Aerosols are not responsible for long-term climate change. Over a period of time, precipitation removes them.

Major volcanic eruptions can have a noticeable short-term effect on climate. For instance, in 1815, the Tambora volcano in Indonesia erupted. A major eruption, its particulates were carried high into the atmosphere and carried around the Earth. Effects were noticed as far away as New England in the United States, which had a "year without a summer." Climatologists calculated that the Tambora eruption alone was responsible for lowering global temperatures by as much as 5°F (3°C). A massive volcanic eruption can cool the Earth's climate for one to two years.

Volcanoes do emit CO_2 and, because this is a greenhouse gas, if enough is released, it can contribute toward global warming. According

Mount St. Helens erupted on May 18, 1980. The eruption sent volcanic ash, steam, water, and debris to a height of 60,000 feet (18,288 m). The volcano lost 1,300 feet (396 m) of altitude and about two-thirds of a cubic mile of material. *(Austin Post, USGS)*

to scientists at the USGS Volcano Hazards Program, there is evidence that the majority of the past 400 million years has seen an elevation in global temperatures because of volcanic eruptions. This ties in with the plate tectonic theory presented in chapter 3.

Just as short-wave radiation comes from the Sun, long-wave radiation comes from the Earth. Like the Sun, the Earth also gives off radiation; but unlike the Sun, the wavelengths given off are very long. It gives off longer wavelengths because it is a much cooler body than the Sun (the hotter the body, the shorter the wavelength). Sometimes it is possible to indirectly see heat radiation. On a hot day, the shimmering effect seen just above a hot road surface is the heat being radiated from the road. The amount of energy given off by the Earth is equal to the amount it receives from the Sun. Otherwise, the Earth would just continue to heat up. Fortunately for life on Earth, the Earth is able to maintain an energy balance, which will be discussed later in this chapter.

Outgoing energy from the Earth comes from several sources: plants, animals, volcanoes, rocks, buildings, and roads, to name a few. All of the outgoing energy added together is the same amount as the amount of radiation the Earth absorbs. This is why the Earth maintains a fairly constant temperature. If the input/output were not equal, then if outgoing radiation was higher than what was absorbed, the Earth would cool down. If outgoing was less, the Earth would heat up.

The ability of an item to give off energy is called *emissivity*. Each object on Earth has its unique emissivity. For comparisons, scientists assign a black object an emissivity of 1.0. They compare all other objects to this base. For example, a rock has an emissivity of 0.8, which means it emits only 80 percent of the absorbed energy compared to what a black object would. The concept of emissivity is important when dealing with global warming because when objects begin to retain—or absorb—heat, instead of emit—or release—it, it heats up the atmosphere. An example of this is CO_2 absorbing infrared energy, which in turn heats the atmosphere.

The general emissivity in areas can also be changed and upset in other ways, sometimes creating a negative result toward increasing global warming. Changing the use of the land's surface can have a significant impact. For example, deforestation of the Earth's rain forests changes the energy balance and the amount of energy leaving the Earth, which can then have long-term effects on the climate.

ATMOSPHERIC ENERGY

There is a tremendous amount of energy flow in the atmosphere. Before any of the Sun's energy can reach the Earth, it must first pass through the atmosphere. Once it enters the atmosphere, several things can happen to it. Some of the energy is directly absorbed by water vapor and ozone. Depending on existing cloud cover, some of it may be reflected directly back to outer space because clouds have a high albedo.

Energy is transferred throughout the atmosphere and to the Earth by three basic processes: radiation, conduction, and convection.

Radiation

Radiation is the direct transfer of heat energy. Energy travels by electromagnetic waves from the Sun to the Earth. The shorter the wavelength, the higher the energy associated with it, such as UV radiation. Con-

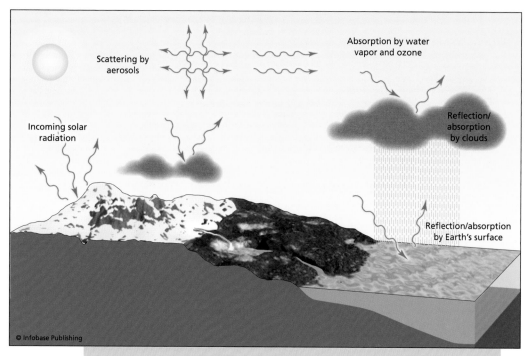

Energy coming from the Sun can be scattered, absorbed, and reflected.

versely, the longer the wavelength, the lower the energy associated with it, such as radio waves and microwaves. The majority of radiation from the Sun is in the visible and near-visible portion of the electromagnetic spectrum. Electromagnetic radiation is what can be reflected, scattered, redirected, and absorbed once it reaches the Earth's atmosphere.

Conduction

Conduction is the process where heat energy is transmitted by being in contact with other molecules. In the process of conduction, an object heats by being in physical contact with another hot object. Each object on Earth is a conductor of some sort. Good conductors are those that responded quickly and heat up when heat is applied to them. Metals are an example of a good conductor. This is why metal cooking pans often have wooden or plastic handles. If the handle were metal, it would burn anyone's hand who tried to pick it up off a hot stove. Conversely, some

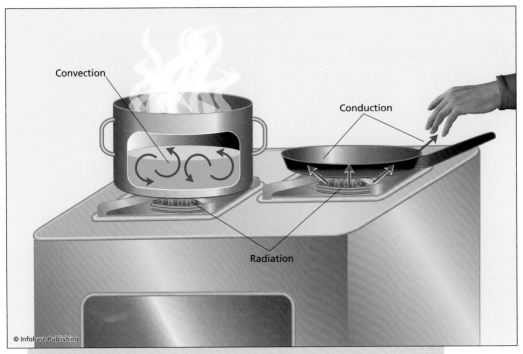

Energy (heat) is transferred between the Earth's surface and the atmosphere by radiation, conduction, and convection.

items do not heat up rapidly when heat is applied to them, making them poor conductors. Water, wood, and some plastics are examples of poor or moderate conductors.

Air is not a very good conductor. Because of this, most of the energy transfer in the atmosphere involving conduction occurs near the Earth's surface. Because the land heats up quickly during the day and cools off quickly at night, it conducts its heat into the air above it. During the day, when the sunshine heats the land, the land heats the air above it. At night, the ground cools quickly, and because energy travels from hot to cold, it draws heat away from the air above the ground, cooling the air.

Convection

Convection heats by transporting entire groups of molecules from one place to another with a substance. The substance is usually a fluid that can move freely, such as water or air. Think of a pot of thick soup on a stove. By conduction, the soup at the bottom of the pan heats first. Then it begins to rise (because heat rises) to the top of the column of soup. As it rises, cooler soup up above sinks to the bottom to take its place. That portion then heats and rises. Soon, a circular pattern of heating has begun in the pot, heating the soup throughout. This is convection.

Convection also occurs in the atmosphere. It can include both large- and small-scale rising and sinking of air masses. These vertical

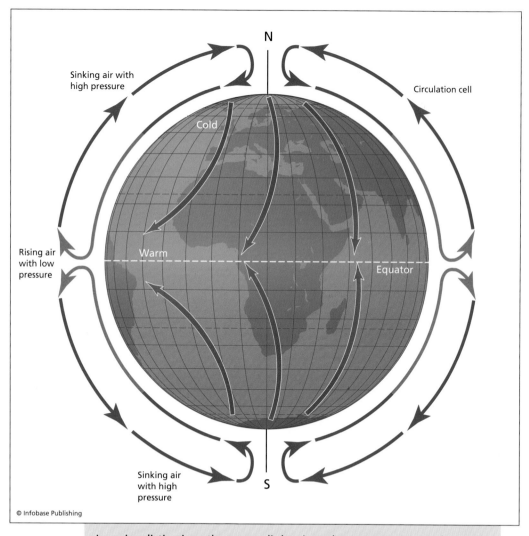

Sinking air with high pressure

Circulation cell

Cold

Rising air with low pressure

Warm

Equator

Sinking air with high pressure

N

S

© Infobase Publishing

In a simplistic view, the warm, light air at the equator rises and spreads northward and southward, and the cool dense air at the poles sinks and spreads toward the equator, forming two convection cells.

motions can be large enough that they are able to actually distribute heat and moisture throughout the entire atmospheric column. This motion and transfer of heat energy is what allows clouds to form and storms to develop.

Storm systems are one scale of convection. There are also very large scales of convection that are able to effectively distribute heat all over the surface of the Earth. The Earth's landmasses, ocean distributions, rotation, and other features can make this somewhat complicated, but to understand the basic flow of convected energy in the atmosphere, think of the Earth in a more simplistic way, such as in the pot of soup example. Most of the Sun's heat is deposited in the Tropics of the Earth. This is because the Earth's rotational axis is an almost perpendicular plane to the plane of the Earth's orbit around the Sun. The polar latitudes receive much less solar heating than the equator. Incoming sunlight warms the equator more than it does both the North and South Poles. This makes the air at the equator warm faster and begin to rise. As it rises, it diverges, and some travels to the North Pole and some to the South Pole.

Because the air at the North and South Poles is cooler and more dense, it sinks and is forced to move toward the equator. At this point, two basic continuous convection cells have begun: one traveling from the equator to the North Pole and back to the equator again in an endless loop; and another one just like it but traveling from the equator to the South Pole and back again. Things become a little more complicated when the rotation of the Earth is added in. Solar heating of the Earth and its atmosphere drives the large-scale atmospheric circulation patterns, and even the seasons. This will be looked at in more detail in chapters 5 and 6.

GLOBAL ENERGY BALANCE

Interactions between energy from the Sun, the Earth, and the atmosphere all have an effect on the Earth. This is called the global energy balance, an energy balance that also plays a role in climate. Because this energy balance has changed and, as a result, the atmosphere is retaining more heat, the process of global warming is already in full swing.

The global energy balance regulates the state of the Earth's climate. Modifications to it—*forcings*—can be by natural sources or human sources, and could cause global climate to change. Natural forcings might include variations in the Sun's intensity, a shift in the Earth's orbit around the Sun, a shift in the Earth's tilt, or an increase in volcanic activity. Human-influenced forcings could include burning fossil fuels, changing land use patterns, or deforestation.

Greenhouse gases in the atmosphere have an effect on the global energy balance. Without the natural greenhouse gases in the atmosphere, the Earth would be uninhabitable because it would be too cold. Several things have an effect on the energy balance, such as clouds and atmospheric aerosols. Clouds can interact in several ways with energy. They can block much of the incoming sunlight and reflect it back to space. In this way, they have a cooling effect. Clouds also act like greenhouse gases and block the emission of heat to space and keep the Earth from releasing its absorbed solar energy. The altitude of the cloud in the atmosphere also has an effect. High clouds are colder and can absorb more surface-emitted heat in the atmosphere; yet they do not emit much heat to space because they are so cold. Clouds can cool or warm the Earth, depending on how many clouds there are, how thick they are, and how high they are. It is not yet fully understood what effect clouds will have on surface temperatures if global warming continues into the next century and beyond.

Climatologists have proposed different opinions. Some think that clouds may help to decrease the effects of global warming by increasing cloud cover, increasing thickness, or by decreasing in altitude. Others think that clouds could act to increase the warming on Earth if the opposite conditions occurred. According to Anthony Del Genio of NASA's Goddard Institute for Space Studies, when air temperatures are higher, clouds are thinner and less capable of reflecting sunlight, which increases temperatures on Earth, exacerbating global warming. Del Genio, in an interview with CNN, said that "In the larger context of the global warming debate I'd say we should not look for clouds to get us out of this mess. This is just one aspect of clouds, but this is the part people assumed would make global warming less severe."

One way that scientists try to predict future climate change and the effects of global warming is through the analysis and interpretation of mathematical climate models. In these computer models, climatologists attempt to account for all items that affect climate. Cloud cover is one of those variables. Today, this is still one of the most difficult variables to control and interpret. The climate is so sensitive as to how clouds might change, that even the most complicated, precise models developed today often vary in their global warming prediction under all the different methods available for cloud modeling.

The main reason clouds are so difficult to model is because they are so unpredictable. They can form rapidly and complete their life cycle in a matter of hours. Other climate variables work on a much slower timescale. Clouds also occur in a relatively small geographic area. Other climate variables operate on a much larger scale. According to climate research scientists at NASA, the world's fastest supercomputers can only track a single column of the surface and the atmosphere every 50 to 200 miles (80–322 km). In comparison, a massive thunderstorm system might cover only 20 miles (32 km). Features that are small, fast, and short-lived are hard to predict. This is one of the reasons why predicting specific individual weather events is more difficult than predicting long-term climate changes over broad areas.

Clouds are just one thing that can change the global energy balance. Snow and ice can also do that. If the Earth becomes cold enough, allowing large amounts of snow and ice to form, then more of the Sun's energy will be directly reflected back to space because snow and ice have a high albedo. Over a period of time, this will change the global energy balance and the global temperature. Conversely, if the Earth warms, the snow and ice will melt. This lowers the surface albedo, allowing more sunlight to be absorbed, which will warm the Earth more.

Deforestation can also upset the global energy balance in many ways. If forested areas are removed and land is left bare, the ground can then reflect more sunlight back to space, causing a net cooling effect. On the other hand, if the forest material is burned, then the carbon stored in the trees is released into the atmosphere, contributing to global warming. Also, since forests are good reservoirs of existing carbon, storing and holding the carbon and keeping it out of the air, if the forest is burned, not only does the already stored carbon now enter the atmosphere, but any future storage potential of carbon in that forest is now destroyed—creating a double addition of carbon toward global warming.

Atmospheric aerosols can be added to the atmosphere by sources such as fossil fuels, biomass burning, and industrial pollution. Tiny smoke particles (aerosols) can either cool or warm the atmospheric temperature depending on how much solar radiation they absorb versus how much they scatter back to space. Fossil fuel aerosols can also pollute clouds. Scientists need to do a lot more research on aerosols before they fully understand the full impacts of aerosols on global warming. The composition of the aerosol, its absorptive properties, the size of the aerosol particles, the number of particles, and how high they are in the atmosphere all have an effect on whether they cool or warm the atmospheric temperature and by how much.

Another effect aerosols have on clouds is that as aerosols increase, the water in the clouds gets spread over more particles, and smaller particles fall more slowly. This could decrease the amount of rainfall. Scientists believe aerosols have the potential to change the frequency of cloud occurrences, cloud thickness, and amount of rainfall in a region. Like clouds, aerosols are also a challenge to accommodate in climate models because they occur over small areas, move rapidly, form and dissolve quickly, and interact with other variables (such as wind) in unexpected ways, making them a hard variable to control.

RATES OF CHANGE

Causes of climate change often trigger additional changes (feedbacks) within the climate system that can amplify or subdue the climate's initial response to them. For instance, if changes in the Earth's orbit trigger an interglacial (warm) period, increasing CO_2 may amplify the warming by enhancing the greenhouse effect. When temperatures get cooler, CO_2 enters the oceans, and the atmosphere becomes cooler. According to the Intergovernmental Panel on Climate Change (IPCC) in 2007, during the past 650,000 years, the CO_2 levels have tended to track the glacial cycles. In other words, during warm interglacial periods, CO_2 levels have been high and during cool glacial periods, CO_2 levels have been low.

Sometimes the Earth's climate seems to be quite stable; other times it seems to have periods of rapid change. According to the U.S. Environmental Protection Agency, interglacial climates (such as the climate today) tend to be more stable than cooler, glacial climates. Abrupt, or rapid, climate changes often occur between glacial and interglacial periods.

There are many components in a climate system, such as the atmosphere, the Earth's surface, the ocean surface, vegetation, sea ice, mountain glaciers, deep ocean, and ice sheets. All of these components affect, and are affected by, the climate. They all have different response times, however. Some are fast, others slow, as shown in the following table.

Climate System Components and Response Times		
CLIMATE SYSTEM COMPONENT	RESPONSE TIME	EXAMPLE
Fast Responses		
Land surface	Hours to months	Heating of the Earth's surface
Ocean surface	Days to months	Afternoon heating of the water's surface
Atmosphere	Hours to weeks	Daily heating; winter inversions
Sea ice	Weeks to years	Early summer breakup
Vegetation	Hours to centuries	Growth of trees in a rain forest
Slow Responses		
Ice sheets	100–10,000 years	Advances of ice sheets over Greenland
Mountain glaciers	10–100 years	Loss of glaciers in Glacier National Park
Deep ocean	100–1,500 years	Deep-water replacement

Because the components of the climate system are diverse in location, function, and size, the way they respond can be diverse as well. The amount of change applied and the innate ability to respond determine what the climate actually ends up doing. For instance, if there is a slow climate change, but the system component reacts quickly, then the response will be visible. If the climate change is rapid, but then reverts back to its previous condition and the component's response time is naturally slow, then there will be no response. If the climate change alternates from one extreme to another at a rate that the components can keep up with, these changes will be seen as visible adaptations. It is these types of rates of change that are most enlightening for climatologists because it allows them to more efficiently model all the subtle components of the climate system.

Planetary and Global Motions in the Atmosphere That Affect Climate

Because the Earth is in motion rotating on its axis and revolving around the Sun, climatologists must also take this into account when they study climate and global warming. In addition, there are global motions in the atmosphere—large-scale features—that have a significant effect on the Earth's local, regional, and global climate. This chapter discusses those planetary motions—eccentricity, tilt, and precession—and how they have been tied to climate change, as well as the role of the Earth's rotation and its influence on the movement of the global atmospheric system. The chapter also addresses the Coriolis force, the trade winds, Hadley cells, monsoon systems, and the El Niño phenomenon.

ORBITAL VARIATIONS

Astronomers have known for centuries that characteristics of the Earth's orbit are not constant; they vary in a cyclic manner. The orbit varies because of the mass gravitational attractions among Earth, the Moon, the Sun, and other planets. These changing gravitational attractions cause variations in the Earth's angle, position, and path; they even slightly alter the positions of the seasons. Slight variations in the Earth's orbit can lead to changes in the distribution and amount of sunlight reaching the Earth's surface.

There are three aspects of the Earth's orbit that change periodically and have an influence on climate—eccentricity, tilt, and precession. Not only do they vary throughout time but they all vary in unique cycles, or periodicity, and the interaction of the cycles also affects the climate. These orbital variations are called the Milankovitch cycles after Milutin Milankovitch, a Serbian astrophysicist.

The Milankovitch cycles are also associated with glaciation and the Earth's past glacial and interglacial cycles. In fact, glacial episodes are related to the cyclical changes in the Earth's orbit around the Sun. Cyclic variations in the Earth's eccentricity, axial tilt, and precession is what creates differences over time in the seasons and the amount and intensity of sunlight reaching various parts of the Earth's surface. During some parts of the cycle, the Earth receives increased amounts of solar radiation; at other times it receives decreased amounts. Climatologists believe this plays a significant part in influencing the advance and retreat of glaciers because it impacts the seasonality and location of incoming solar energy.

Eccentricity

Eccentricity is the shape of the Earth's orbit around the Sun. Its orbit is not perfectly circular, but rather slightly oval in shape and the degree of ovalness changes throughout time. Because it is oval in shape, its orbit changes the distance from the Earth to the Sun at different times of the year. Now, the Earth is the closest to the Sun in its orbit in January and farthest away in July (about 3 percent difference). Currently the Earth's orbit is only slightly oval—which means the eccentricity is low. When the eccentricity increases and the orbit becomes more oval, the difference could became 20 percent more energy reaching the Earth when it is closest to the Sun. This could have a significant effect on heating the atmosphere.

The Earth's eccentricity occurs in a cycle that has been well documented. Climatologists refer to this cycle as periodicity. The periodicity is 100,000 years—that is, the time necessary to change the orbit from a nearly circular one to a more elongated one. It also turns out that the Earth's ice ages also peak about one every 100,000 years and have for the past million years.

Axial Tilt

Axial tilt, or the tilt of the Earth's axis, is the second of the Milankovitch cycles. This is the inclination of the Earth's axis in relation to its plane of orbit around the Sun. If the Earth were not tilted, there would be no seasons.

The seasons are created by the change in length of daylight hours. As the seasons progress, the daylight hours get shorter (in winter) or longer (in summer) and the noon Sun changes its altitude in the sky (high altitude in the summer, low in the winter). Because the angle of the Sun changes, it affects the amount of solar energy the Earth receives. When the Sun is directly overhead (summer months) it only has to pass through the thickness of one atmosphere. But when the rays enter at a lower angle (such as 20 degrees), they must pass through two atmospheres. If the Sun is only 5 degrees above the horizon (very low), it is the equivalent of having to pass through 11 atmospheres, meaning that very little solar energy reaches the Earth's surface. The more atmosphere the energy must pass through, the more likely it is that it will be scattered, reflected, and absorbed before reaching the Earth's surface.

The changes in the Sun's angle and the length of day are a direct function of the Earth's axial tilt. Currently, it is tilted 23.5 degrees from the perpendicular. The tilt is referred to as the inclination. On the summer solstice (June 21), the North Pole is inclined 23.5 degrees toward the Sun. This is the Northern Hemisphere's first day of summer (the Northern Hemisphere receives the most light and energy from the Sun) and the Southern Hemisphere's first day of winter. Six months later on the winter solstice (December 22), the North Pole is inclined 23.5 degrees *away* from the Sun. This is the Northern Hemisphere's first day of winter (the Northern Hemisphere receives the least light and energy from the Sun) and the Southern Hemisphere's first day of summer.

The Earth's tilt does not remain constant at 23.5 degrees. It can vary from 21.5 to 24.5 degrees. When the tilt is less (toward 22.5 degrees), the seasons will vary less. Remember, if there were no tilt to the axis, there would be no seasons at all. Less of a tilt causes the Sun's radiation to be more evenly distributed between summer and winter. It also increases the difference in the amount of radiation reaching the equator versus the poles. Scientists propose that when the Earth's axial tilt is less, it could change the climate and promote the growth of ice sheets. They reason that winters would be warmer, allowing the air to hold more moisture, which would produce more snowfall. Summer temperatures would be cooler, so less of the increased winter snowfall would melt. Conversely, when the Earth's axial tilt moves in the other direction and reaches the 24.5 degree range, the opposite conditions would apply. The oscillations in the Earth's axial tilt occur on a periodicity of 41,000 years from 21.5 to 24.5 degrees.

Precession

The third of the Milankovitch cycles is precession. Precession is the Earth's slow wobble as it spins on its axis. While the Earth rotates on its axis, it is not a perfect rotation. Like a spinning top winding down begins to wobble as it spins on its axis, the Earth does the same. The wobble slowly changes the direction toward where the Earth's axis is pointed. Today, for example, the Earth's north axis is pointed at Polaris, the North Star, but that has not always been the case, nor will it always remain the case. The precession of the Earth wobbles from pointing to Polaris to pointing at the star Vega. When this shift occurs, Vega is the Earth's new North Star. The Earth's wobble, or precession, has a periodicity of 23,000 years. This is important to the Earth's climate because unlike now where the Earth is closest to the Sun during the winter solstice, when Vega becomes the North Star, the Earth will be farthest away from the Sun during the winter solstice and will be closest to the Sun during the summer solstice. This will cause greater seasonal contrasts in the climate.

The Milankovitch cycles provide a theory of the potential influence of these factors on the occurrence of ice ages throughout time. Climatologists have found evidence of these orbital forcings in climatic data—the evidence of the 100,000-year eccentricity cycle being the strongest. Scientists believe they are significant because, although they do not affect the total solar energy received by the Earth, they do affect where and during which season the sunshine is received, which can ultimately affect climate. Much research is still needed in order to determine how big a role these orbital forcings have on global warming.

THE EARTH'S ROTATION AND THE CORIOLIS FORCE

When pressure differences alone are responsible for moving air, the air—or wind—will be pushed in a straight path. For example, if someone opens a door into an enclosed space containing a different air pressure, they will feel a rush of air when the door is opened and the air travels from higher pressure to lower pressure. It may seem odd when looking at satellite weather maps to see that large storm clouds move in circular patterns. Winds do follow curved paths across the Earth. Named after the scientist who discovered this effect, Gustave-Gaspard de Coriolis, this phenomenon is called the *Coriolis force*. The Coriolis force is an apparent drifting sideways (*deflection*) of a freely moving object as seen by an observer on Earth. It is the tendency for any moving body on or above the Earth's surface, such as an ocean current, an air mass, or a ballistic missile, to drift sideways from its course because of the Earth's *rotation* underneath. In other words, a moving object appears to veer from its original path. In the Northern Hemisphere, the deflection is to the right of the motion; in the Southern Hemisphere, it is to the left.

The Coriolis deflection of a body moving toward the north or south results from the fact that the Earth's surface is rotating eastward at greater speed near the equator than near the poles, since a point on the equator traces out a larger circle per day than a point on another latitude nearer either the North or South Pole (the equator is a great circle and other latitudes are smaller). A body traveling toward the equator with the slower rotational speed of higher latitudes tends to fall behind or veer to the west relative to the more rapidly rotating Earth below it

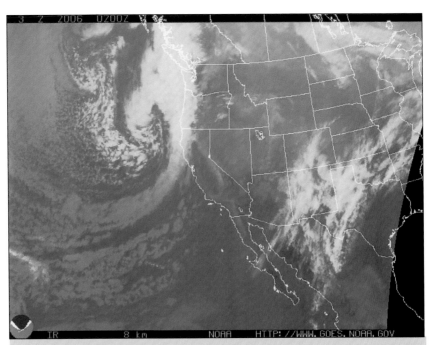

The Coriolis force can be seen each night in satellite weather photos as the clouds take on their characteristic circular formations. *(NOAA)*

at lower latitudes. Similarly, a body traveling toward either the North or South Pole veers eastward because it retains the greater eastward rotational speed of the lower latitudes as it passes over the more slowly rotating Earth closer to the pole.

The practical applications of the Coriolis force are important when calculating terrestrial wind systems and ocean currents. Scientists studying the weather, ocean dynamics, and other related Earth phenomena must take the Coriolis force into account.

The Coriolis force is the reason that tropical storms or cyclones rotate counterclockwise in the Northern Hemisphere (called hurricanes) and clockwise in the Southern Hemisphere (called typhoons). The reason why a hurricane does not form on the equator is because the equator has the weakest Coriolis force. A hurricane forms just to the north of the equator and generally heads northwest if not interfered with by other winds or pressures. It is also the Coriolis force that affects dominant wind patterns, such as the westerlies, easterlies, horse latitudes, and equatorial doldrums. The Coriolis also affects regional and global weather patterns because it interacts with the jet streams.

LARGE-SCALE FEATURES OF THE ATMOSPHERIC WINDS

Large-scale circulation in the Earth's atmosphere transports heat from low to high latitudes. Over some zones around the Earth, the winds

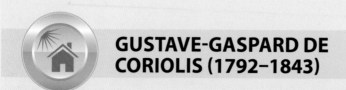

GUSTAVE-GASPARD DE CORIOLIS (1792–1843)

Born in Paris on May 21, 1792, Coriolis was a French engineer and mathematician who first described the force of motion on a rotating body, which has come to be called the Coriolis force. It is the apparent path of deflection of an object as it passes over the rotating Earth. If an object is launched, it will not land in a straight trajectory but be offset depending on the rate of movement of the object and the rate of movement of the Earth underneath. This concept was very important to the fields of meteorology, ballistics, and oceanography.

Coriolis showed that if the ordinary Newtonian laws of motion of bodies are to be used in a rotating frame of reference, an inertial force, acting to the right of the direction of body motion for counterclockwise rotation or to the left for clockwise rotation, must be included in the equations of motion. The effect of the Coriolis force is an apparent deflection of the path of an object that moves within a rotating coordinate system. In reality, the object does not actually deviate from its path, but it looks like it does because of the motion of the coordinate system.

When referring to the surface of the Earth, the Coriolis force is the most apparent in the path of an object moving longitudinally (from pole to pole) because this is where the greatest velocity and movement take place. On the Earth, an object that travels along a north-south path— or longitudinal line—will undergo apparent deflection to the right in the Northern Hemisphere and to the left in the Southern Hemisphere, because as an object is traveling through the air, the Earth is constantly moving under the object. When the object lands, it is not at the point expected directly north (or south), but its path looks curved because of the deflection.

blow predominantly in one direction throughout the year and are associated with the rotation of the Earth. Over other areas of the Earth, the dominant wind direction changes with the seasons. Around the equator there is a belt of relatively low pressure, called the *doldrums*, where the heated air expands and rises.

Tropical heating fuels a huge tropical circulation pattern called the Hadley cell. Here, warm air rises in giant columns. This is the region

The other reason for the phenomenon is that the tangential velocity of a point on the Earth is a function of latitude (the velocity is basically 0 at the poles and it attains its maximum value at the equator). To illustrate this concept, think of a line of figure skaters in an ice show. If the line traces out a circle with one end as the pivot (not traveling across the ice, but turning in place), and the rest of the line rotates around it, a skater in the middle of the line will not have to skate nearly as fast as the skater on the far end. That skater will need to skate faster than the others because they have extra ground to cover in the same amount of time. As another example, if a cannon were fired northward from a point on the equator, the projectile would land to the east of its due north path. This variation would occur because the projectile was moving eastward faster at the equator than was its target farther north. Likewise, if the weapon were fired toward the equator from the North Pole, the projectile would again land to the right of its true path. This time, the target area would have moved eastward before the shell reached it because of its greater eastward velocity. A similar displacement occurs no matter the location from which the projectile is fired.

The Coriolis force is very significant in astrophysics and stellar dynamics. It is used to study the rotation of sunspots. It is also significant in Earth science, especially meteorology, physical geology, and oceanography because the Earth is a rotating frame of reference and the forces working on it affect everything else in the system. The Coriolis force is extremely important in the study of the dynamics of the atmosphere, such as the prevailing winds and the rotation of storms, and it has an effect on the ocean currents. Coriolis contributed immensely to many scientific disciplines.

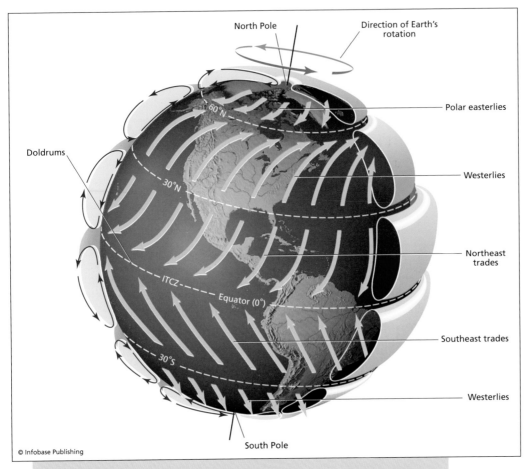

© Infobase Publishing

The general circulation of the atmosphere: Heated air rises in the Tropics at the ITCZ and sinks in the Subtropics (30°). Additional heat flows toward the polar areas, while cold air circulates from the poles toward the equator. This transfer of heat creates major circulation cells that form the major wind systems, such as the trade winds, westerlies, and polar easterlies. These circulation patterns will be altered as global warming intensifies.

where towering cumulonimbus clouds created by the evaporation of water vapor from the tropical oceans occurs. The enormous amounts of water vapor are able to condense into the thick, puffy cloud formations common in the humid Tropics. This condensation produces a narrow zone of rainfall in the rising part of the Hadley cell near the equator. Huge amounts of heat are transported through the Hadley cell.

As the air rises and loses water vapor in the Tropics, it moves toward the subtropics in both hemispheres (a Hadley cell moves northward in the Northern Hemisphere and southward in the Southern Hemisphere). Once it reaches the subtropics, at a latitude of 30 degrees (north and south, respectively) the air in the cell cools and sinks back toward Earth. The sinking air is then warmed by the increasing pressure of the atmosphere at lower elevations, where it becomes even drier and able to hold more water vapor.

The Hadley cell flow prevents condensation from occurring in the subtropics, making these latitudes (30°N and 30°S) a zone of low precipitation and high evaporation. It is in this area that the Sahara is located. At this point, the *trade winds* from both the Northern and Southern Hemispheres blow from the subtropics back toward the Tropics and replace the rising air at the equator once again. As the warm, dry air is carried by the trade winds over the tropical ocean, it is able to pick up water vapor from the ocean's surface. The region near the equator where the northern and southern trade winds meet is called the Intertropical Convergence Zone (ITCZ). The ITCZ is a zone of abundant rainfall, due largely to the water vapor that the trade winds gather off the ocean on their way to the equator.

The Hadley cell is an extremely important component of the poleward transfer of heat to the Earth's polar areas. These large-scale movements of air also determine the pressure (the weight of the air) at the Earth's surface. When air moves up and away from the Tropics, it reduces the weight of the air mass over the ITCZ and causes a low surface pressure. The air that moves downward into the subtropics (30° latitude) has a higher surface pressure because the weight of air is pressing downward.

Because solar heating is the driving force behind the Hadley cell circulation, the seasonal shifts of the Sun between the hemispheres also affect the location of the ITCZ. It moves northward during the Northern Hemisphere's summer and southward during the Northern Hemisphere's winter (Southern Hemisphere's summer). The slow thermal response of the land and oceans causes the seasonal shifts of the ITCZ to lag about a month.

Important seasonal transfers of heat between the tropical ocean and land are called *monsoons* and happen because water takes much longer

DISCOVERY OF THE TRADE WINDS

Christopher Columbus (1451–1506), a Genoese seaman and explorer, is credited with the discovery of the trade winds. It was the trade winds that he used in order to sail his three ships across the Atlantic Ocean. He was able to sail from the Canary Islands to the Bahamas in 1492. An amazing accomplishment for the time, it was a distance of 5,400 miles (8,690 km), and he completed his voyage in only 36 days due to the force of the trade winds.

Much later, in 1970, Thor Heyerdahl, a Norwegian seaman and archaeologist, sailed a ship made of reeds from Morocco to the Caribbean, illustrating that the trade winds were capable of helping sailors accomplish this feat. He proposed that because it was possible to use the trade winds to accomplish this, that perhaps the idea of building pyramids could have been spread to Central America and Mesoamerica by Egyptians who may have traveled this same route long ago.

For centuries, the trade winds have been known to mariners as being consistent enough to provide transcontinental trade routes. They are the most steady, consistent wind systems on Earth. The expression *the wind blows trade* comes from navigating by the trade winds and means "the winds blows on track."

to heat up and cool down than land does. A monsoon is a major wind system that changes direction on a seasonal basis. In the summer monsoon, air flows from over the cooler ocean toward the land, where it quickly heats, rises, condenses, and produces heavy precipitation and releases huge amounts of heat. The strongest summer monsoon area in the world is in India where there is a strong, wet summer monsoon against the Himalaya Mountains. The winter monsoon is the reverse of the summer monsoon. The cold, dry air flows down and out over the land toward the ocean and precipitation occurs over the ocean.

Although monsoons are seasonal—giving areas a rainy season and a dry season—they have lasting effects on climate and affect many areas of the world, such as Asia, Africa, North America, and South America.

In the subtropical regions at 30° north and south, high pressure dominates. Also in this area the influence of the monsoonal flow of air from land to sea in summer produces oval-shaped cells of high pressure over the subtropical oceans. Air naturally flows away from higher pressure toward lower pressure, but, as the atmosphere circulates out of the high-pressure cells, it gets deflected by the Coriolis force. This creates the winds called the *westerlies*. The westerlies are a surface flow of warm air out of the subtropics that transport heat to the cooler high latitudes.

In the higher middle latitudes in both hemispheres, the circulation in the lower atmosphere is a complex zone of transition between the warm air flowing out of the subtropics and the cold flow from the polar latitudes toward the equator. The weather in this zone is dynamic, with a wide variety of climate conditions. The constant creation of high- and low-pressure cells that move from west to east (the westerlies) are separated by frontal zones. Frontal zones are regions near the Earth's surface where large changes in temperature occur over small areas due to fast-moving air. The poleward movement of warm air and the equatorward movement of cold air along the frontal systems that these provide serve to warm the polar latitudes and help maintain the Earth's energy budget. These midlatitude cyclones will be covered in chapter 6.

GLOBAL WARMING AND ATMOSPHERIC CIRCULATION

According to a study conducted in 2006 by Gabriel Vecchi of the University Corporation for Atmospheric Research, the trade winds in the Pacific Ocean are weakening as a result of global warming. This conclusion is based on the findings of a study that showed the biology in the area may be changing, which could be harmful to marine life and have the long-term effect of disrupting the marine food chain. Researchers predict that it could also reduce the biological productivity of the Pacific Ocean, which could have an impact on not only the natural ecosystem and balance, but also the food supply for millions of people.

The study used climate data consisting of sea-level atmospheric pressure over the past 150 years and combined that with computer

Much of the world's population depends on fish as a food source. Not only El Niño, but also global warming and climate change have negative effects on marine life. *(NOAA, Fisheries Collection)*

modeling to conclude that the wind has weakened by about 3.5 percent since the mid-1800s. The researchers predict another 10 percent decrease is possible by the end of the 21st century.

Some of the computer modeling simulations included variables such as the effects of human greenhouse gas emissions, while other simulations included only natural factors that affect climate such as volcanic eruptions and solar variations. Vecchi concluded that the observed weakening of the trade winds could only be accounted for through the model that included human activity—specifically from greenhouse gases and the burning of fossil fuels. According to an interview on the LiveScience Web site, Vecchi believes "this is evidence supporting global warming and also evidence of our ability to make reasonable predictions of at least the large scale changes that we should expect from global warming."

Vecchi believes global warming is to blame because in order for the ocean and atmosphere to maintain an energy balance, the rate that the atmosphere absorbs water from the ocean must equal the amount that it loses to rainfall. As global warming increases the air temperature, more water evaporates from the ocean into the air. The atmosphere cannot convert it to rainfall and return it back fast enough. Because the air is gaining water faster than it can release it, it gets overloaded and the natural system compensates by slowing the trade winds down, decreasing the amount of water being drawn up into the atmosphere, in order to maintain the energy balance.

THE ORIGIN OF EL NIÑO'S NAME

El Niño was named by Peruvian fishermen who first noticed a warm current that appeared off the coast of South America at the beginning of the calendar year during the Christmas season. El Niño literally means the "little boy" or the "Christ child" in Spanish.

It was not until the 1960s that El Niño was recognized as a global phenomenon that effected changes over the entire Pacific. Today, it is studied and recognized as a physical process with a huge influence on the climatic conditions over large areas of the world.

The name El Niño refers to the warm phase of a larger oscillation in which the surface temperature of the central and eastern part of the tropical Pacific varies up to 6.7°F (4°C) and is associated with changes in the winds and rainfall patterns. The complete, larger phenomenon is called the El Niño–Southern Oscillation (ENSO). The warm El Niño phase lasts for eight to 10 months. The entire ENSO cycle lasts about three to seven years. The Pacific Ocean signatures are important temperature fluctuations in surface waters of the tropical eastern Pacific Ocean. ENSO is associated with floods, droughts, and other climatic disturbances. ENSO has signatures in the Pacific, Atlantic, and Indian Oceans. While it is a natural part of the Earth's climate, scientists today are concerned about whether or not its intensity or frequency may change as a result of global warming. The first official description of ENSO in terms of its physical properties and behaviors was by Dr. Jacob Bjerknes of the University of California, Los Angeles, in 1969.

On January 10, 2005, a landslide hit the community of La Conchita in Ventura County, California, destroying or seriously damaging 36 homes and killing 10 people. Landslides and debris flows are one of the consequences of El Niño. *(Randy Jibson, USGS)*

The drop in winds reduces the strength of both the surface and subsurface ocean currents and interferes with the cold-water upwelling at the equator that is responsible for supplying ocean ecosystems with valuable nutrients. This could seriously hurt all marine life.

This is an example of how global warming could affect atmospheric circulation, which would then affect ecosystems over large areas. Global warming can affect other atmospheric circulation patterns as well, such as the seasonal monsoon systems. Much of the world's population relies on the monsoon rains. If that balance was upset, it could cause disastrous effects on food and water supplies, ecosystems, health issues, and more. The sometimes obvious, and other times subtle, effects of global warming are serious and far-reaching.

EL NIÑO

The El Niño-Southern Oscillation (ENSO) phenomenon is a global mechanism that is caused by the large-scale interaction between the ocean and the atmosphere. The Southern Oscillation (a more recent discovery) refers to an oscillation in the surface pressure (atmospheric mass) between the southeastern tropical Pacific and the Australian-Indonesian regions. When the waters in the eastern Pacific are abnormally warm—referred to as an El Niño event—sea level pressure drops in the eastern Pacific and rises in the west. The reduction in the pressure gradient is accompanied by a weakening of the low-latitude easterly trades. This enables warm water near Australia and Indonesia (the eastern Pacific) to travel eastward toward Peru and Ecuador (South America). Normally, the trade winds move warm water westward, piling warm surface water in the west Pacific so that the sea surface is roughly 1.6 feet (0.5 m) higher at Indonesia than at Ecuador. The sea surface temperature is about 13°F (8°C) higher in the western Pacific than the east. This causes the upward movement of deeper, colder water to the surface off the coast of South America. This is called upwelling, and it is an extremely important process because it

Large hail collects on streets and grass during a severe thunderstorm. The larger ice stones are two to three inches (5–7 cm) in diameter. These storms can be very destructive to homes, cars, and yards. *(NOAA, National Severe Storms Laboratory)*

Storms can cause tornadoes as a result of El Niño. This photo shows a tornado touching ground in Cordell, Oklahoma, on May 22, 1981. *(NOAA)*

brings up rich nutrients that make it a very rich, productive fishing area. The upwelling process feeds the plankton, which in turn feed the fish. El Niño, however, spreads warm water eastward *toward* South America and it covers the cold upwellings with about 500 feet (152 m) of water keeping it warmer than average, which plays havoc with marine ecosystems. The Southern Oscillation (the SO in ENSO) is the atmospheric component of the cycle.

El Niño has widespread impacts on a global scale. The area of warm water, and its associated energy, is enormous. Flooding, damage to ecosystems, increased spreading of infectious diseases, drought, wildfires, crop failures, and starvation are some effects of El Niño. It also causes increased tropical storms and hurricane activity.

In 1997, Hurricane Linda, which made landfall on the coast of Mexico, was a result of El Niño. It was one of the fiercest eastern Pacific storms ever recorded, with winds that reached 185 miles per hour (298 km/hr). El Niño can also cause intense rainfall and mudslides. California is especially vulnerable to mudslides and debris slides triggered by El Niño. Because many of California's urban areas are built on steep, coastal mountain slopes, when the soil becomes saturated, the entire slope can fail, bringing entire houses down.

Ice storms are another extreme weather effect from El Niño. When warm air flows over a freezing surface, a heavy layer of ice can form. The ice can become so heavy when it is deposited on surfaces that it can cause extreme damage. It can break trees, damage homes, and bring down power lines, leaving homes without electricity for long periods of time.

El Niño has also been credited with causing tornadoes across many areas of the United States. In 1998, Florida suffered the worst series of tornadoes in the state's recorded history. It was reported that 800 homes were completely destroyed and 700 were left uninhabitable. Forty-two people lost their lives and damages exceeded $60 million.

El Niño can also cause a die-off of marine life or reduce the survival rate of the young. It has also been known to devastate popula-

Changes in the conditions of marine systems can endanger the survival rate of sea lion pups. *(Captain Budd Christman, NOAA)*

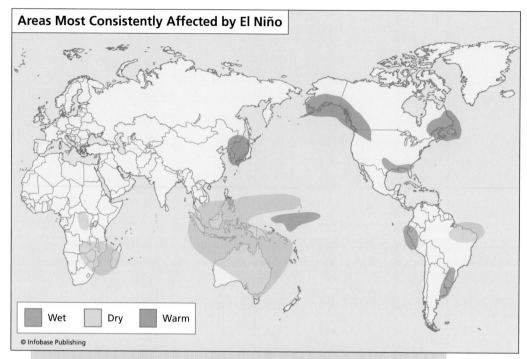

Areas Most Consistently Affected by El Niño

Wet   Dry   Warm

© Infobase Publishing

Specific regions will be affected by El Niño. Some areas will have an excess of rain, others drought. Some areas will be warmer than usual, others cooler.

tions of seabirds off Peru by reducing the fish stock that they feed off. Some scientists believe that if El Niños were to become more frequent the overall composition of marine life would change.

Today, there is some debate in the scientific community as to whether there is a connection between El Niño and global warming. According to the National Oceanic and Atmospheric Administration (NOAA), some climatologists say that the series of warm climate events during the 1990s are evidence that a general warming trend is starting to change the weather; yet others say that the variations are within the normal limits. One of the major problems at this point is that scientists have not been documenting El Niño events long enough—in order to draw reliable scientific conclusions, they need to be able to document events over a long period of time, the longer the better (hundreds to thousands of years, or more).

El Niño Time Line	
YEAR OF OCCURRENCE	COMMENTS
1567–68	First El Niño event historically recorded
1630–31	
1641	
1650	
1694–95	
1715–16	
1782–83	Stronger than previous occurrences
1790–93	
1802–04	
1823–33	Multiple El Niños occurred during this time interval
1844–46	This El Niño was stronger than previous occurrences
1951–52	
1953	
1957–58	This was the first El Niño to receive recognition by scientists worldwide
1963–64	
1965–66	
1968–70	
1972–73	Extremely strong El Nino whose climatic effects were felt worldwide in the form of drought, flooding, and other responses

(continues)

(continued)

El Niño Time Line	
1976–77	
1982–83	This El Niño event went down in the history books as the largest one on record
1990–92	This El Niño event was comparable in intensity to that of 1982–83. During this series, the United States experienced one of the warmest winters on record
1993	Very strong, similar to 1982–83
1994–95	Very strong, similar to 1982–83
1997–98	This El Niño event was the strongest one in the 20th century
2002–03	Very strong
2004–05	Very strong
2006–07	Very strong
Source: NOAA	

Even so, there is a fairly large group of experts at NOAA, National Aeronautics and Space Administration (NASA), and the U.S. Geological Survey (USGS) who believe that El Niño events have become not only more frequent, but more intense during the past 100 years. During the 20th century, there were 23 El Niños. The four strongest of these all occurred after 1980. According to the Intergovernmental Panel on Climate Change (IPCC), the frequency and intensity of El Niño events have increased since 1970 compared with the previous 100 years. They state that since 1976 El Niños have become more frequent, persistent, and intense.

Although there is still debate about whether global warming causes El Niño, it is accepted that El Niños do produce more heat, which is carried out of the Tropics and raises global temperatures. Scientists predict that if global warming increases El Niños will also increase.

Computer models have been an extremely powerful tool in helping climatologists predict climatic conditions. By inputting field-collected data along with physical laws of the ocean and atmospheric general circulation patterns, systems can be modeled.

Models are built on the basis of present observations. Then, in order to check their reliability, they are tested using past real-world observations to see if they can derive the same results and match what happened in reality. Once this has been successfully achieved, the models can then be used to predict the future. Models to predict El Niño are currently being improved so that they can predict an upcoming event a year or more in advance.

Local Motions in the Atmosphere That Affect Weather and Climate

Local motions in the atmosphere also affect the weather and climate of an area. These patterns play a large role in determining what plants and animals can live in the region. The ecosystem of any area has grown up around its climate and, if that climate is altered, every species in the ecosystem must either adapt, migrate, or die. One of the major concerns about global warming is that it will affect ecosystems, upsetting delicate balances between climate, plants, and animals. If major food chains are upset, it can have far-reaching, devastating impacts. In order to understand what effects global warming can have on local environments it is necessary to understand local atmospheric systems. This chapter discusses regional wind systems and the effect of continentality, the jet stream, and midlatitude cyclones. It presents local wind systems, such as sea-land breezes, monsoon winds, orographic uplift, lake effect, and mountain-valley breezes. Finally, it covers extreme weather events, such as tropical cyclones.

REGIONAL WIND SYSTEMS

Differences in atmospheric pressure are what cause the wind to blow. Pressure differences—also called pressure gradients—can happen at many scales. They can be global or local, but no matter what size, they develop because of differences in heating and cooling of the Earth's surface. It is the heating/cooling cycles that happen on a daily basis that cause many of the local wind systems.

Air flows from high pressure to low pressure. This pressure gradient is set up when an area of land receives more sunshine than another area and begins to heat up faster. As evidenced by a hot air balloon, warm air rises. The Earth's surface warms the air directly above it through the process of conduction and convection. This forms an area of high pressure, which will then naturally flow to an area of low pressure, because a pressure gradient has been set up. An upper air low pressure may exist because of the presence of clouds, keeping the ground beneath it cooler than the section of ground beneath the high pressure.

As the upper air high pressure air flows to the upper air low pressure, this air then sinks to the Earth's surface, which creates a high pressure on the ground. Along the ground, this high pressure air flows to the low pressure area under the upper air high, where it is lifted, creating a circulation cell as the air travels in a circular pattern from one pressure gradient to another as shown in the illustration on page 96.

The presence of land also makes a difference in weather patterns and seasons. Seasons are much more greatly defined over land than they are over the oceans. Because of this, there is a significant difference between the seasonality of climate in the Earth's Northern Hemisphere than in the Southern Hemisphere. In the Northern Hemisphere, where most of the Earth's landmasses are located today, weather differences are much more dramatic. This is because land heats up and cools down more rapidly than water, making the resulting temperature ranges much greater. Thus, the impacts of global warming will be felt more strongly in the Northern Hemisphere.

(continues on page 98)

There are various atmospheric circulation patterns that affect local and regional wind systems, such as (a) high and low pressure systems as a result of heating the Earth's surface; (b) development of a sea breeze due to the more rapid heating of land during the day; (c) development of a land breeze due to the more rapid cooling of land during the night;

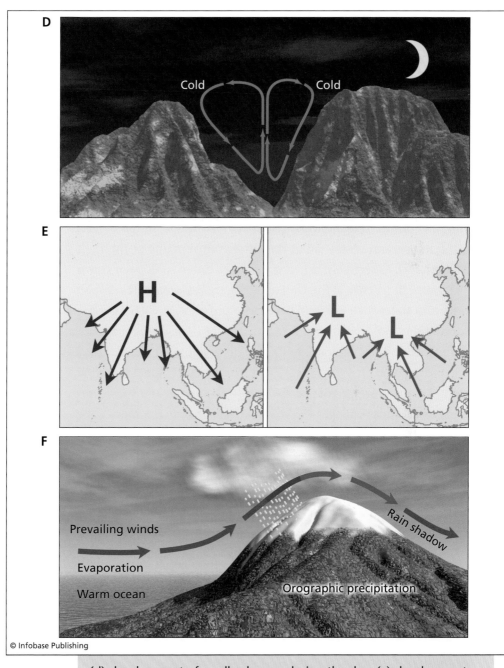

(d) development of a valley breeze during the day; (e) development of a mountain breeze at night; (f) typical monsoon patterns of winter and summer creating a pronounced dry and wet season; and (g) orographic precipitation, creating a windward and leeward side of the mountain with distinct vegetation patterns. *(modeled from USGS)*

(continued from page 95)

The *jet stream* refers to a channel of fast-flowing westerly air that resides high in the atmosphere—at a height of 20,000 feet (6,096 m). Usually existing in the stratosphere, it forms at the boundaries of warm and cold air masses. There are two principal jet streams—one in the polar region in each hemisphere between 30° and 70°N. These wind channels extend over long distances around the Earth. Sometimes they can split. They play a critical role in determining the weather because they steer storms and control the positions of high and low pressure regions. These winds can travel 50 knots (58 mph) or faster and exert a significant control over the weather in North America. In the winter, when the temperatures stay mild, the jet stream remains at the high (polar) latitudes. Conversely, when it becomes bitter cold, even down into the southern reaches of the United States, the jet stream has dipped far south into states such as Utah, Colorado, and New Mexico. In some cases, variations in the jet stream bring on greater than average rainfall, which causes flooding. Climatologists also believe that jet streams play a role in the creation of storms that spawn tornadoes. If global warming affects the temperature gradients on the Earth's surface and the atmosphere, it could cause changes in the jet streams that could change the weather in local or regional areas.

Midlatitude cyclones are the cause of most of the storms in the midlatitudes—such as the United States. These are the storm systems that bring snowy blizzards, flooding rains, lightning storms, and other severe types of weather to the latitudes between 30° to 60°N. They are huge moving systems of low pressure caused by the interaction of warm tropical and cold polar air.

These cyclones (rotating storm systems) generally move toward the east. It is these weather systems that are most commonly carried by the jet stream. Therefore, by understanding the movement of the jet stream, climatologists will be able to predict the weather for an area.

Midlatitude cyclones can generate a wide variety of storms, such as hail, sleet, freezing rain, and rain, snow, and ice pellets. Precipitation is most common at the center of the low pressure and along the fronts where the air is being quickly uplifted. A slower warm front of air usually leads, followed by a faster-moving cold front. When the cold front catches up with the warm front (both pivoting around the low pressure) the warm air mixes, a process called occlusion. The warm, uplifted air condenses and forms cumulus, then cumulonimbus clouds, which can develop into severe thunderstorms.

The occurrence of severe weather is directly related to global warming. Scientists believe continued warming will cause increases in severe weather events. Although these storms do not typically cause as much damage as tropical cyclones (hurricanes) they can still cause severe damage from wind destruction and flooding. Global warming can alter weather patterns, change surface climate, change atmospheric circulation, and change extreme events. According to the IPCC in their 2007 global warming assessment report, cyclone activity over both hemispheres has changed since 1950. General features include a poleward shift in storm track location, increased storm intensity, and a decrease in total storm numbers. Some scientists have reported that the North Atlantic storm track has shifted 112 miles (180 km) northward in the winter, which could be related to reduced midlatitude winter precipitation. If this is because of global warming, it could permanently affect ecosystems. The IPCC has also documented a gradual reduction of the number of days with frost and an increase in the number of warm nights in the midlatitudes in recent decades.

LOCAL WIND SYSTEMS

There are several types of local wind systems: sea-land breeze, lake breeze, monsoon, orographic precipitation, lake effect, and mountain-valley breeze.

Sea-Land Breeze

A sea-land breeze is a thermal circulation system that develops along shorelines. These breezes are created where there are variations in pressure due to differences in the way both land and water cool. During the day, the land heats faster than the water does under the sunlight, which makes heated air rise. The air over the oceans is a cool high because water does not heat nearly as fast as land. As soon as a big enough difference is reached and there is a significant enough pressure gradient, air will begin to flow from the cooler ocean air to the land, replacing the air rising in the thermal low. This, in turn, causes the heavier, cooler air over the ocean to move toward land, replacing the air rising in the thermal low. This circulation cell is referred to as a *sea breeze*, because the wind is flowing in from the ocean. A sea breeze usually does not start until midmorning, because the surface of the Earth needs time to heat up and set up the pressure gradient, which triggers the cellular flow. After several hours of heating, sea breezes are strongest by late afternoon, because this is when the greatest temperature and pressure contrasts exist.

Once the Sun sets and the heat source is removed, the land begins to cool down quickly (much quicker than the water, because water retains its heat longer), until both the land and ocean are similar in temperature again. At sunset, the opposite process occurs. During the night, the land cools off more quickly than the ocean, and soon a large enough temperature gradient has again been set up. This time, however, the opposite circulation pattern happens. The land surface is cooler than the water, making it the thermal high pressure area. The ocean now becomes the warm thermal low. The wind now moves from the land to the ocean and is called a *land breeze*.

Lake Breeze

A lake-land breeze is similar to the land-sea breeze and can develop as a local wind circulation pattern. In this system, the inland moving wind is called a lake breeze. Lake breezes are common in late spring and summer along the Great Lakes region of the United States.

Monsoon

A monsoon is a regional-scale wind that changes directions on a seasonal basis. Similar to sea-land breezes, monsoons are also caused by temperature contrasts, though on a much larger scale. Their wind flow also corresponds with the seasons. During the summer season, the continents heat up much faster than the oceans, causing the warm, moist air from the ocean to blow in from the ocean over land, creating periods of heavy rainfall. As the warm, humid air blows on shore, the moisture is condensed. This is the wet season.

The winter months have the opposite effect. The ocean surface is now warmer than the land, enabling the winter monsoons to bring clear, dry weather and winds that blow from the land out across the ocean. Monsoon circulation patterns are common in India, Australia, Africa, South America, and North America. Increases in global warming could alter these major systems, affecting areas worldwide through drought or flooding.

Orographic Precipitation

Another local motion in the atmosphere is caused by the presence of a mountain range and the resulting precipitation pattern that occurs when prevailing winds are consistently pushed over a specific side of the mountain barrier. In the midlatitudes, when low-pressure cells move eastward they interact with the topography. This is common in the western United States when air masses encounter the massive Rockies. Air flow eastward from over the Pacific Ocean carries huge amounts of water vapor. When this air encounters mountain ranges, its forward movement is blocked. The air mass is then forced to rise to higher elevations to get over the mountain range.

As the air rises, it cools. Water vapor then condenses from the cooling air and produces heavy rainfall on the side of the mountain that the air was forced up—the windward side. This is called orographic precipitation. Once the air reaches the crest of the mountain and begins moving down the other side, most of the water vapor has been depleted, making the air very dry. As the air sinks down the mountainside, it is compressed and warmed. At this point, the air can store more water vapor without condensation occurring. Because of this, the lee or rain shadow side of mountain ranges is an area of lower precipitation and can look much different than the windward side. This is one of the factors that causes the interior areas of the United States to be dry. If global warming alters the dynamics of moisture availability and wind systems, this could cause further dryness.

Lake Effect

Another local weather phenomenon is called the lake effect. Lake effect is an episode of heavy winter snowfall that can occur when a mass of cold Arctic air moves over a body of warmer water. This creates an unstable system. The storm pulls moisture from the body of water, which freezes

The windward side of a slope in an orographic wind flow pattern receives the greater share of the moisture from an air mass being pushed up and over a mountain barrier. Vegetation on this side is lusher. *(Nature's Images)*

and is deposited as snow on the lee side of the lake. In some cases, when the enriched air mass is pushed up a mountain after crossing a warm water body, the orographic effect can enhance the lake effect snowstorm. The orographic component is able to produce intense amounts of snowfall. If the temperature is not low enough to form snow, it will form heavy rainstorms. Lake effect snow is a weather phenomenon in southeast Canada, the northeast United States, and the Great Salt Lake in Utah. The amount of precipitation an area will get from lake effect snow is determined by several factors—the direction of the wind, its duration, and the magnitude of the temperature difference between the water and air. In fact, the lake effect of the Great Salt Lake in Utah is one of the factors that provides snow each winter to support recreational skiing. If global warming reduces snowfall amounts, severe economic effects will be felt over large regions in transportation, entertainment, recreation, lodging, food, and other industries.

Mountain-Valley Breezes

Mountain-valley breezes are common in areas with significant topographic relief. During the day, as the Sun heats up the land and air at the valley bottom and sides, a *valley breeze* develops. When the air heats, it becomes less dense and more buoyant, allowing it to start flowing up the valley sides. The vertical rise of the air along the sides of the mountain is restricted by a temperature inversion layer, confining the airflow to the valley (not letting the rising air escape into the atmosphere). Once the rising air pushes against the inversion, it moves horizontally toward the center of the valley, and drops toward the valley floor. Like the other local systems illustrated in this chapter, it creates a self-contained circulation system. Oftentimes, this cycle of air will develop cumuliform clouds at the mountain peaks.

During the night, the opposite happens. The air along the mountain slopes cools quickly. As it cools, the air becomes denser and begins to

The leeward side of a slope in an orographic wind flow pattern receives little moisture from an air mass being pushed up and over a mountain barrier. Vegetation on this side is drier and scarcer. *(Nature's Images)*

flow downhill, causing a *mountain breeze*. It converges on the valley floor and forces the air to move vertically upward. This upward movement is usually stopped by the temperature inversion. Similar to daytime flow, it forces the air to move horizontally, which then allows the airflow to complete the cycle in a cell. Where canyons and valleys are narrow, they can funnel the winds; some winds have reached speeds as high as 93 miles per hour (150 km/hr). Mountain-valley breezes are important for firefighters to understand when they battle wildfires, which are one of the consequences of global warming (global warming leads to drought and excessive evaporation, which dries out the exist-

A STORM WITH MANY NAMES

Hurricane actually comes from the name Huracan, the god of evil to the Tainos tribe from Central America. Other parts of the world use different terms for these storms. In the western Pacific and China Sea, they are called typhoons. *Typhoon* comes from the Cantonese word "tai-fung," which means "great wind." In countries such as Pakistan, India, Australia, and Bangladesh, they are called cyclones. In the Philippine Islands they are called baguios. The true scientific term for the storm is tropical cyclone. A cyclone is simply a very large system of rotating air that pivots around a point of low pressure. It is only the tropical cyclones, which have warm air in the centers, that have the potential to become the powerful, destructive storms we know as hurricanes.

According to Christopher Landsea, the Science and Operations Officer at the National Hurricane Center of the National Oceanic and Atmospheric Administration (NOAA), the following all mean hurricane:

Foracan	hericane	uracano	heuricane
harauncana	hurlicano	furacane	hyrricano
herican	uracan	harrycain	haracana
hurlecan	duracana	herocane	haurachana
oraucan	haroucana	urycan	hurleblast
foracane	hericano	furicano	jimmycane
haraucane	hyrracano	hauracane	

ing vegetation, making it vulnerable to wildfire from lightning strikes). Because the winds change direction, firefighters must plan their strategy accordingly, so that they are prepared for abrupt changes in the path of the fire.

EXTREME WEATHER

Global warming raises serious concerns over its potential to cause damage to people, property, and the environment as a result of extreme weather events, such as severe drought and storms. Today, scientists, such as Dr. Christopher Landsea at NOAA, are trying to understand just how much impact global warming may have on the occurrence and frequency of drought, hurricanes, and tornadoes. It still remains difficult to assess because global warming will have different impacts on different areas of the Earth. Although they cannot predict exactly where hurricanes or other severe storms will occur in the future, they are fairly certain that as the atmosphere continues to warm under the influence of global warming, it will cause an increase in heat waves. Because warmer air can hold more moisture, it will change the hydrologic cycle, which will alter flooding and drought patterns. There is a great concern that increasing ocean temperatures will also increase the likelihood of tropical cyclones—or hurricanes.

The World Meteorological Organization (WMO) has warned that extreme weather events, such as drought, hurricanes, and heavy rainfall, may very likely increase because of global warming. They have noted that episodes of extreme weather have been on the rise in recent years and that the rate of events is increasing. In fact, in a report released in 2002, they concluded that the severe drought Australia suffered that year was due principally to human-induced global warming.

There are not many forces in nature that can compare to the destructive capability of a hurricane. These storms can have winds blow for long periods of time at 155 miles per hour (249 km/hr) or higher. Not only is the wind destructive, but also the rainfall and storm surges can cause significant damage and loss of life.

Hurricane Katrina, which formed August 23, 2005, and dissipated August 30, 2005, affecting the Bahamas, South Florida, Cuba, Louisiana, Mississippi, Alabama, and the Florida Panhandle, was the deadliest hurricane in the history of the United States, killing more than 1,800 people and destroying more than 200,000 homes. It created more than 900,000 evacuees and was the costliest hurricane in U.S. history, with more than $75 billion in estimated damages. Today, only about 40 percent of New Orleans pre-Katrina residents have returned to the city.

A hurricane, or tropical cyclone, forms over tropical waters—between latitudes 8° and 20° in areas of high humidity and light winds, where the sea surface is warm. Typically, temperatures must be 80°F (26.5°C) or warmer for a hurricane to start. This is why global warming and the heating of the ocean are such concerns.

The most typical time for a hurricane to form is in the summer and early fall of the tropical North Atlantic and North Pacific Oceans, making the Northern Hemisphere's hurricane season run June through November.

NASA'S EXTRATROPICAL STORM TRACKS ATLAS

Climate scientists at the National Aeronautics and Space Administration (NASA) have developed a database available online that displays the extratropical storm tracks that occurred between 1961 and 1998, allowing visitors to find out what the weather was like on the day they were born or any other day in the featured time span.

The original purpose for the creation of the climate atlas was to look back and try to assess the impact of global warming on storms. The online atlas plots the paths of storms and records statistics by tracking the atmospheric low-pressure centers at sea level. The system records several pieces of information, such as storm frequency and intensity, along with the paths of individual storms. It also produces maps of average storm intensity of the most severe storms. The site provides data on individual monthly and seasonal averages for the years 1961 to 1998 as global maps in a variety of projections. Besides images, the hard data is also available for downloading, which contains information on length and location of storms, atmospheric pressure at the storm's center, day, month, year, and time. NASA is currently working to expand the years of coverage to go back to 1950 and extend through 2001.

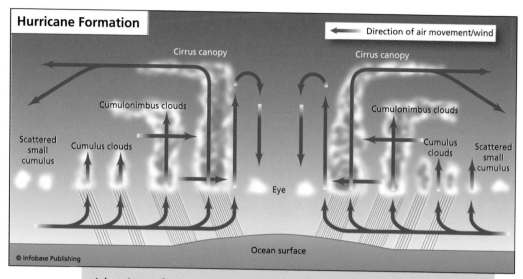

A hurricane forms when the movement of warm, humid air and cold, unstable air between the upper and lower atmospheres meet. Strong, gusty winds and heavy rains fed by the warm ocean water enter the hurricane from the bottom.

A hurricane begins with the creation of a tight group of thunderstorms over the tropical oceans. Meteorologists refer to this condition as a *tropical disturbance*. These disturbances start when warm winds converge (collide). When the air collides, it forces it to move upward, starting a thunderstorm.

Often, when the trade winds of the Northern and Southern Hemispheres meet in the Intertropical Convergence Zone (ITCZ), they collide and rise. If a group of thunderstorms breaks away from the ITCZ, it can form an organized storm system. Storms can also form if warm and cold air collide. Once the storm gets organized, the surface air pressure falls in the area around the storm and winds begin to spin in a cyclonic circulation (counterclockwise in the Northern Hemisphere, clockwise in the Southern Hemisphere). Latent heat gets released, which makes the air rise, while the surrounding air sinks. The sinking air gets compressed and warmed. At the surface, it is drawn back toward the center and rises. This process sets up a circular cell. Increasing winds spin around the center of the storm and draw heat and moisture from the warm ocean surface, which provides more fuel to keep the system

Tropical cyclones, such as this one off the coast of Brazil, are some of the most deadly storms on Earth. *(NOAA)*

going. This system becomes self-perpetuating. A cycle of evaporation and condensation brings the ocean's heat energy into the vortex (center) and fuels the storm.

The storm takes the easily recognized spiral shape because of the Coriolis force generated by the rotation of the Earth. Once a storm system starts and its winds reach 23 miles per hour (37 km/hr), its category moves from a tropical disturbance to a *tropical depression*. When winds increase to 39 miles per hour (63 km/hr), the cyclone is referred to as a tropical storm and receives an official name. As soon as the winds reach 74 miles per hour (119 km/hr), the storm system is classified as a hurricane. A hurricane can only persist if conditions in the upper atmosphere remain just right. If high pressure exists in the upper atmosphere, it keeps the air currents that are needed to fuel the hurricane from rising. Also, if strong upper level winds exist, they can keep a hurricane from forming by ripping the thunderstorms apart, which keep temperatures from warming and make it impossible for a low pressure system to form at the surface, effectively cutting off the energy to fuel the hurricane. A hurricane will also fall apart if it moves over cooler water with no supply of warm, moist air, or if it moves over land.

Protecting people and the environment from severe weather triggered by global warming currently has many research scientists at the National Hurricane Center at NOAA and elsewhere engaged in theoretical studies, computer modeling, and collection and analysis of field data in an effort to gain a better understanding of the mechanics of global warming and its interaction with the environment to improve forecasting, response, and safety.

Ocean Currents

In order to fully understand global warming, it is necessary to understand the world's ocean currents, where they are, how they work, how they interact with the atmosphere, how they influence climate on land, and how they interact with the Earth's heat budget. This chapter will examine the basic principles of ocean circulation and the roles of seawater density, temperature, and salinity. It will then discuss specific ocean phenomena such as the Intertropical Convergence Zone (ITCZ), the Pacific Decadal Oscillation (PDO), the North Atlantic Oscillation (NAO), and the Great Ocean Conveyor Belt. Finally, it will explain what the potential long-term consequences could be if these delicate balances are upset.

OCEAN CIRCULATION

At the surface, ocean currents are driven by the winds, which make the surface water move parallel to the predominant wind direction, except

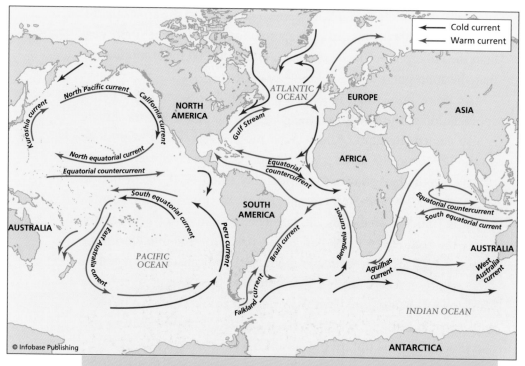

The surface ocean circulation transports heat globally, maintaining temperatures and supporting marine life. The principal warm and cool currents control climate locally, regionally, and globally.

where continental landmasses exist and block its movement. Water also moves vertically in the ocean depths. There are two factors that make water more dense (which causes it to sink) or less dense (which keeps it on the surface): (1) salt content and (2) temperature.

Atmospheric flow and ocean currents are the mechanisms that carry heat from the equator to the poles. There are many processes that can alter the circulation patterns, which can change the weather of an area. Variations can happen on cycles that last months, years, decades, or centuries. Variations can also affect deepwater or surface water. If the ocean did not distribute heat throughout the world, the equator would be much warmer and the poles would be much colder.

The oceans are equally important as the atmosphere in transporting heat from the equator to the polar regions. In terms of how much heat and water it can hold, its capacity is much greater than the atmosphere's. In fact, the world's oceans can store approximately 1,100 times more heat than the atmosphere. The oceans also contain 90,000 times more water than the atmosphere does.

As more knowledge is gained about global warming, a better appreciation of the role the oceans play in shaping the Earth's climate is also gained. Because of this, much more research has been done on the oceans in the past 15 years, leading to the discovery that the oceans' depths have warmed considerably since 1950. According to scientists at the Woods Hole Oceanographic Institution, until recently, scientific models predicting climate change could not account for where the projected warmth had gone—it was unaccounted for in the atmosphere. This discrepancy in the model had caused much confusion until researchers finally figured out that the missing greenhouse warming was being stored by the world's oceans. Water has a tremendous capacity to hold heat. The warming had occurred, but no one had thought to look toward the oceans for the answer.

Now that scientists understand this relationship, those who study global warming agree that including the ocean system in global warming studies is critical. Not only do the oceans have an enormous ability to hold heat, the constant movement of slow and fast water can affect the weather for months at a time. It is important for climatologists to understand these ocean interactions in order to be able to predict regional trends in climate. It is also important to understand deepwater processes and processes that occur near the surface to understand what mechanisms drive climate.

The oceans also play a critical role in balancing the carbon dioxide (CO_2) levels. The CO_2 levels in both the atmosphere and the ocean reach an equilibrium. If something happens to upset this balance—such as changes in chemistry—then sudden shifts in the CO_2 levels can affect the climate. This is one of the concerns about the steadily increasing levels of greenhouse gases. If the oceans reach the point where they cannot absorb CO_2, it could upset the balance of ocean currents and climate patterns on a global scale.

THE ROLES OF VARIABLE SEAWATER DENSITY, TEMPERATURE, AND SALINITY

In the world's oceans, the properties of density, temperature, and salinity (salt content) all work together and result in distinct characteristics that ultimately relate to climate change and global warming. Solar energy is absorbed by seawater and stored as heat in the oceans. Some of the energy that is absorbed may evaporate seawater, which increases its temperature and salinity. When a substance is heated, it expands and its density is lowered. Conversely, when a substance is cooled, its density increases. The addition or subtraction of salts also causes seawater density to change. Water that has higher salinity will be denser.

Pressure is another factor that affects density. Pressure increases with depth; this can be felt when a swimmer goes underwater and has to relieve ear pressure at certain levels. As pressure increases with depth, so does the density of a water mass. Because high density water sinks and low density seawater rises, this distinct change in density generates water motion. This concept is extremely important in the world's oceans because it is a chief mechanism controlling the movement of major currents and ocean circulation patterns.

Oceanographers and climatologists are interested in the distributions of both temperature and salinity in the world's oceans because they are two factors that determine the vertical thermohaline circulation (see chapter 1). Thermohaline comes from two words: *thermo* for heat and *haline* for salt. Of the three factors—temperature, salinity, and pressure—that have an effect on water density, temperature changes have the greatest effect. In the ocean, the thermocline (a water layer within which temperature decreases rapidly with depth) acts as a density barrier to vertical circulation. This layer lies at the bottom of the low density, warm surface layer and the top of the cold, dense bottom waters. The thermocline keeps most of the ocean water from being able to vertically mix because these two layers are so drastically different. In the polar regions the surface waters are much colder than they are anywhere else on Earth. This means they are denser, so that little temperature variation exists between the surface waters and the deeper waters—basically eliminating the thermocline. Because there is no thermocline barrier, vertical circulation can take place as the surface waters sink (a process called *downwelling*), where they replenish deep waters in the major oceans.

Water surface temperatures have significant effects on coastal climates. Because seawater can absorb large amounts of heat, it enables coastal locations to have cooler temperatures in the summer than inland areas. Coastal currents also affect local climate. For example, Los Angeles, California, and Phoenix, Arizona, are at similar latitudes, yet Los Angeles has a much more moderate summer climate because of the effect of the ocean.

Another influence on surface temperatures is a phenomenon called *upwelling*. Upwelling is the rising of cooler waters from greater ocean depths. In some coastal locations, the motion of the wind along with the Coriolis force pushes the surface water offshore, allowing deeper water to rise from below and replace it. Upwelling is common in areas along the Florida, Oregon, and California coasts. These are the areas where rising cool, nutrient-rich water supports commercial and sport fishing in the waters just offshore. These areas are important to the economies of the local regions.

Because water of high density sinks and water of low density rises, the change in density is an extremely important process involved in the movement of water—the creation of currents. Scientists study both the distribution of salinity and temperature of seawater to track ocean currents. Surface water sinks because its temperature and salt content changes as it moves along the surface and it becomes denser than the water beneath it. The water that sinks does not mix with the surrounding water, enabling the sinking water masses to be identified from computer data displays that show temperature, salinity, and nutrient cross sections of the deep ocean.

Another important factor to consider in the circulation of seawater is the mechanism involved in the formation of sea ice. Freshwater freezes at 32°F (0°C). The addition of salt lowers its freezing point. At a salinity of 35 percent, the freezing point of water is lowered to 28.8°F (-1.91°C)—roughly 3°F (2°C) lower than freshwater. This is the reason why many cities salt their roads in the winter during snowstorms to keep the roads clear of snow and ice.

As a result, in polar regions when freezing occurs salt is physically released from the ice as it forms, which further lowers the freezing temperature of the nearby water. This is because the water next to the ice becomes saltier and denser, and new freezing stops until the temperature drops to the new freezing point. Density, temperature, and salinity are important to the major circulation patterns of the world's oceans.

THE INTERTROPICAL CONVERGENCE ZONE

The ITCZ is a belt of low pressure that circles the Earth at the equator. It is formed by the upward flow of warm, humid air from just north and south of the equator. Air is brought into the ITCZ through the force of the Hadley cell and lifted high in the atmosphere by convection. This area is a prime area for thunderstorm activity. Areas in this region receive precipitation more than 200 days each year.

The ITCZ changes position seasonally, shifting north and south of the equator more than 315 miles (500 km). In July, it moves north of the equator, in January it moves south. The increasing Coriolis force makes the formation of tropical cyclones within this zone more likely. The ITCZ has a significant effect on the rainfall patterns of many equatorial countries, including the wet and dry seasons. Because of the ITCZ, these countries commonly experience droughts or flooding episodes.

THE PACIFIC DECADAL OSCILLATION

The PDO is another cyclic pattern in the oceans that causes climate variability. This pattern affects the Pacific Ocean on a cycle that lasts 20 to 30 years. The PDO is most visible in the North Pacific/North American area (north of 20° N latitude). During a warm (positive) phase, the western Pacific becomes cool and the eastern part warms. During a cool (negative) phase, the western Pacific is warm and the eastern cool.

One consequence of the PDO is major changes in northeast Pacific marine ecosystems. When the phenomenon is in a warm cycle, coastal fisheries are fertile and productive in Alaska, but less productive off the west and northwest coast of the contiguous United States. Cold cycles are the opposite. Scientists have characterized the PDO by changes in (1) sea surface temperature, (2) sea level pressure, and (3) wind patterns.

Scientists have not been able to figure out yet what causes the PDO. The 20–30 year phases seem fairly consistent: cool phases existed from 1890 to 1925 and 1945 to 1977. Warm phases existed from 1925 to 1946 and 1977 to 1998. Some scientists, such as Benjamin Giese of the College of Geosciences at Texas A&M, have suggested that long-term changes in the Pacific Ocean temperatures may be an important clue to understanding global warming. Because of the abrupt changes that occur on a 20–30 year cycle in the PDO, they propose that maybe global warming is as much a result of some natural cyclic variation as human activity.

THE NORTH ATLANTIC OSCILLATION

The NAO is the largest, most powerful climate system on Earth. It affects the climate of the Northern Atlantic and the regions that surround it—mainly Europe and the eastern United States. Unlike El Niño, the NAO does not generate violent weather phenomenon; its effects are more widespread than local.

In order to understand the NAO, it is important to understand the concept of high and low pressure. Air pressure is simply a measure of how much air is pushing down at a particular location. Air mass and temperature differences between the Earth's surface and the upper atmosphere create vertical currents. This is what initially creates high- and low-pressure systems.

In a low-pressure system, the weight of the air is low at the Earth's surface. The vertical winds in this system travel *upward,* sucking air upward with it, away from the Earth's surface. Because the air is being pulled upward, it lessens the pressure at the ground. Because it leaves a low void at the Earth's surface, atmospheric currents moving along the Earth's surface are pulled inward from the surrounding areas at the base of the low and spin counterclockwise in the Northern Hemisphere (clockwise in the Southern Hemisphere).

In a high-pressure system, air is pushed down toward the ground, adding pressure from the atmosphere. The air moves vertically downward. This causes the atmospheric currents to spin clockwise in the Northern Hemisphere (counterclockwise in the Southern Hemisphere). The air is pushed away from the center of a high-pressure system.

Both the low- and high-pressure systems can evolve into huge circular rotating systems. The higher in pressure a high-pressure system becomes or the lower in pressure a low-pressure system becomes, the stronger it gets, and the stronger it gets, the more intense the circulation pattern becomes.

The driving mechanism that runs the NAO is a difference in pressures: a high-pressure system over the Azores Islands and a low pressure system over Iceland. Although both systems are present all year long, it is during the winter season that the NAO has a significant effect on climate. During the winter months, both the high pressure and low pressure fluctuate in intensity, and it is the way they fluctuate relative to each other that causes distinctive variations in climate. When the difference between both the high pressure and low pressure is large—both pressures are *strong*—then their effect is to make winters warmer in northern Europe, make it warm in the northeast United States, and cause drought in the Middle East. This is considered a positive NAO.

When the pressure difference between the low and high systems is small—both pressure systems are weak—then they make the Mediterranean countries of Europe rainy and wet, the Scandinavian countries of northern Europe (Finland, Sweden, Norway, Denmark) extremely cold, and the eastern coast of the United States cold. This is a negative NAO.

The NAO oscillates on a cyclic pattern on a decade-to-decade timescale. Scientists have been monitoring it since the 1960s and have determined that the NAO tends to be positive for three to five years and then shifts and becomes negative for three to five years.

The NAO is being studied today by several scientists at National Aeronautics and Space Administration (NASA), as well as the Climate Prediction Center at National Oceanic and Atmospheric Administration (NOAA), trying to understand its behavior: why it behaves as it does, what causes it, what drives it from year to year, whether global warming is involved, and what are the possible short- and long-term consequences. Some scientists believe it may be a response to formation of sea ice or currents in the sea. With supercomputers becoming more capable, these scientists are currently trying to build mathematical models to mimic and predict its behavior.

Even though the NAO has an impact on the entire Atlantic, its largest impact is in Europe where it causes winter storms. When the high over the Azores and the low over Greenland both develop into extremely large, strong circulation patterns, there is a zone that forms between them with a strong-moving current of air that effectively channels the weather systems that pass from the United States directly into Europe. As the pressure systems vary from week to week, they affect the storms, as well as the amount of rain and snowfall that is generated for winter storms by the air passing over the Atlantic and the warm Gulf Stream, determining the amount of precipitation and temperatures experienced in Europe.

When the difference between the high- and low-pressure systems is large (a positive NAO), the strong winds in the channel push winter storms north toward Scandinavia and northern France. Conversely, when the pressure difference is small (a negative NAO), the storms typically travel from the southern United States to southern Europe, the Middle East, and northern Africa.

According to Jim Hurrell at the National Center for Atmospheric Research, the direction these storms take during the winter months can have a dramatic effect on Europe's weather. A positive NAO can make winters warmer by 5°F (3°C). During winters where Europe experiences a consistently positive NAO, it can lengthen the growing season by 20 days in Sweden, provide sunny, warm beaches on the French Riviera, but cause water shortages in the fertile crescent of Egypt. According to NOAA, a long duration of a negative NAO (small pressure difference) does not change southern Europe's rainfall, but significantly cools off northern Europe. The Alps experience superb skiing conditions, and the Mediterranean remains warm, increasing the production of olives and grapes in Greece. Although the most pronounced effects are felt by the European countries, the United States also feels an effect. When the NAO is positive, the eastern United States generally experiences warmer temperatures with less snowfall.

One of the biggest challenges of the NAO for scientists is their ability to predict it. They believe if they could predict a year in advance just how positive or negative the NAO was going to be, they could advise European farmers when to plants crops, advise the ski and tourist industry on predicted snowfall patterns, advise communities of predicted winter temperatures, and advise transportation departments of predicted maintenance workloads. Scientists are currently trying to use computer modeling to find answers to their questions.

THE GREAT OCEAN CONVEYOR BELT

The ocean waters are constantly moving from winds that create waves and the pulling of gravity, which causes the tides. One of the most significant features in the ocean is the thermohaline circulation, commonly referred to as the great ocean conveyor belt. This massive, continuous loop of flow plays a critical role in determining the climate of the world. The two mechanisms that make this conveyor belt work are heat and salt.

The great conveyor belt plays a major part in distributing the Sun's heat around the globe after it has been absorbed by the ocean. In fact, if it wasn't for this flow, the equator would be much hotter, the poles would be much colder, and western Europe would not enjoy as warm a climate as it currently does.

The ocean conveyor belt does not move fast, but it is enormous. It carries 100 times as much water as the Amazon River. The mechanism that drives it is the differences in density. It is the ratios of salt and temperature that determine the density. When water is cold and salty, it is denser and sinks. When it is warm and fresh, it is less dense and rises to the surface.

The conveyor belt literally travels the world. In general, in a continuous loop, it transfers warm water from the Pacific Ocean to the

Atlantic as a shallow current, then returns cold water from the Atlantic to the Pacific as a deep current that flows farther south. Specifically, as it travels past the north of Australia, it is a warm current. It travels around the southern tip of Africa, then moves up into the Atlantic Ocean. At this point, it turns into the Gulf Stream, which is a very warm, north-flowing current critical for providing warmth to western Europe and the northeast coast of the United States. After it passes western Europe and heads to the Arctic, the surface water evaporates and the water cools down, releasing its heat into the atmosphere. It is this released heat that western Europe enjoys as part of their moderate climate.

At this point, the water becomes very cold and increases in salinity, becoming very dense. As its density increases, it begins to sink. The cold, dense water sinks thousands of feet below the surface of the ocean. It is now at extremely deep depths and slowly moves southward through the deep ocean abyss in the Atlantic. It flows to the Southern Hemisphere south of Australia and heads north again, until it eventually mixes upward to the surface of the ocean and starts the process all over. This entire conveyor belt cycle takes about 1,000 years to complete.

Scientists at Argonne National Laboratory (ANL) have been actively researching how long it takes water to move through this conveyor system by tracking levels of the radioactive argon-39 isotope in the Earth's ocean circulation. They are interested in tracking changes in ocean circulation, because scientists have long believed that this system plays an important role in climate moderation. According to Ernst Rehm, a physicist at ANL, "We have some idea that if the 'conveyor belt' stops, then the warm water that is brought to Europe will stop. We have some idea that this may cause an ice age in Europe."

THE CONSEQUENCES OF DESTABILIZATION—ABRUPT CLIMATE CHANGE

The great ocean conveyor belt plays an extremely important role in shaping the Earth's climate. A slight disruption in it could destabilize the current and trigger an abrupt climate change. Climatologists at NOAA and NASA believe that as the Earth's atmosphere continues to heat from the effects of global warming there could be an increase in precipitation as well as an influx of freshwater added to the polar oceans as a result of the rapid melting of glaciers and ice sheets in the Arctic Ocean. They believe that large amounts of freshwater could dilute the Atlantic Gulf Stream to the point where it would no longer sink to the ocean depths to begin its return from the polar latitudes back to the equator.

Measurements taken over the past 40 years have shown that salinity levels within the North Atlantic region are slowly decreasing. What makes this so serious is that if cold water stopped sinking—which means the Gulf Stream would slow and stop—there would be nothing left to push the deep, cold current at the bottom of the Atlantic along, which is what ultimately drives the worldwide ocean current system today.

If this were to happen, the results would be dramatic. Western Europe and the eastern part of North America would cool off. Temperatures could plummet up to 8.3°F (5°C). This is about the same temperature as in the last ice age.

The Global Warming Issue

Global warming is one of the most controversial issues society has ever had to deal with. Because it affects every person on Earth and must be dealt with on a global scale, it has captured the attention of not only scientists, but politicians, academicians, economists, merchants, farmers, planners, medical specialists, engineers, and every other person who has an interest in the environment, food supply, water supply, health, and future of life on Earth. This chapter takes a look at an early view of global warming, how that relates to scientific response in general, why there is such a controversy, and why scale has had such an influence. It then explores how advances in technology and education have played a role in the outlook of the problem and what the growth of public environmental awareness over the past few decades has done. Finally, it touches on the effects of the media's response and public's reaction to the issue.

EARLY STUDIES OF GLOBAL WARMING AND SCIENTIFIC RESPONSE

Global warming was first discovered more than 100 years ago in 1896 by a Swedish chemist named Svante Arrhenius. He claimed that fossil fuel combustion enhanced the global warming effect. Sadly, the scientific community did not take his ideas seriously. The Industrial Revolution was just getting started and there was no overwhelming evidence of increasing atmospheric levels of carbon dioxide (CO_2). The issue held

SVANTE ARRHENIUS (1859–1927): SWEDISH CHEMIST

Svante Arrhenius was the first scientist to predict that the climate was changing because of human activities. With the Industrial Revolution gaining momentum, in 1896 he discovered that the amount of CO_2 in the atmosphere was increasing. He predicted that CO_2 concentrations would increase along with the increase of fossil fuels being used by industry. At that time, coal was the principal fossil fuel used. It was considered a dirty fuel, adding substantial amounts of CO_2 into the atmosphere. He understood that CO_2 heated the atmosphere when fossil fuels were burned.

He predicted that if atmospheric CO_2 levels doubled the Earth would become several degrees warmer. This was the first scientific suggestion of global warming. No one paid any attention. Today, his theory is widely accepted.

Arrhenius was awarded the Nobel Prize in chemistry in 1903, was elected a foreign member of the Royal Society in 1911 and was awarded its Davy medal, and also received the Faraday Medal of the Chemical Society in 1914.

Arrhenius made several noteworthy contributions to science. He was a forward-thinking scientist who had the ability to see complex issues on a global scale. He was also one of the pioneers of environmental science and, although he never lived to see it, set a scientific foundation for the problem of global warming.

no immediate harmful consequences for the public, making the issue relatively unimportant, in spite of Arrhenius's warnings. His studies led him to believe there was a relationship between CO_2 levels in the atmosphere and temperature. Through his calculations, he figured out that if the atmosphere's natural level of CO_2 were doubled, it would cause the temperature to rise by 8.3°F (5°C). Around the same time, Thomas Chamberlin, an American geologist, also hypothesized that humans could warm the Earth by adding CO_2 to the atmosphere. Interestingly, these early conclusions describing global warming were arrived at as a byproduct of other research geared toward explaining the ice ages. The issue at the time was whether lower CO_2 levels could explain the causes of ice ages, whether lower CO_2 levels could cause colder atmospheric temperatures. Conversely, it stood to reason that as CO_2 levels in the atmosphere climbed, the higher the temperature would get.

Even though what Arrhenius and Chamberlin had discovered was important and would one day become a critical issue, it wasn't recognized as such at the time and was soon forgotten.

SCIENTIFIC VIEWPOINTS AND SCALE

Although the human cause of global warming had been presented years before it was accepted as a scientific theory, it was not taken seriously at first due to its scale. At the beginning of the 20th century, many scientists did not think that humans could have a significant effect on the world's environment; they believed any human contribution would be small. Many scientists were researching solar activity (sunspots), ocean circulation, and changes in the Earth's axis, orbit, and tilt. It was generally accepted that any effects caused by humans would be minuscule in comparison with the forces of nature.

Scientists also did not worry much about CO_2 levels, because they believed that the Earth's oceans would act naturally as a giant carbon sink, absorbing whatever amount of CO_2 was added to the atmosphere. There was a consensus that the ocean would effectively absorb all the pollution people could possibly add to the environment.

Another obstacle that has slowed the acceptance of global warming as an urgent issue is the range of scientific viewpoints on the subject. Scientists often do not work in absolutes; they reason in probabilities.

This way, there is always room for change if something new is discovered. They are more comfortable in couching their conclusions in case somebody does more research and discovers something new that disqualifies, discounts, or modifies in some way the previously accepted assessment. This way, if something new is discovered and the end results change, they do not lose credibility. Thus, the 2,500 climate scientists involved in the IPCC research process give their conclusions as a probability calculation rather than a definitive certainty.

Scientists also have differing viewpoints on how they perceive nature and the way it responds to change. One viewpoint is that nature is very predictable and stable and, whatever inputs it receives, it will regain its balance. Followers of this scenario would not be concerned with increases of fossil fuel emissions or other human-induced causes of global warming. They would assume the system will adjust.

A second viewpoint is the exact opposite of the one just presented. In this scenario, nature is highly vulnerable to human impacts, and humans must take great care because unhealthy behavior could destroy it.

These are the extremes. A third viewpoint combines the two. Here, nature is looked upon as resilient and able—most of the time—to tolerate human impacts. This only works to a point, however. If the limits of the natural system are exceeded, the natural system may not be able to recover. In the case of global warming, the Earth may be able to tolerate a certain amount of CO_2 being added to the atmosphere, oceans, soil, and organisms and still be able to adjust and exchange adequately in the carbon cycle. Eventually though it will reach a saturation point and will not be able to maintain an equilibrium. The sink will get too full and will start overflowing; a point of no return.

A last viewpoint, a belief in nature as chaos, completely unpredictable, subscribes to the belief that managing it in any way is pointless. The problem with this theory is that humans take no responsibility for their environment.

Scientists with all of these views determine how global warming is handled. It has been difficult to reach a consensus. There is no prior civilization's written record to compare it to; it is not known what outcomes will be. Human beings today are the guinea pigs taking part in an Earth-sized laboratory experiment. As more information is collected and studied, most scientists agree that humans are changing the environment, adding to the greenhouse effect.

The tables on pages 127, 129, and 130 list scientists' varying viewpoints and illustrate why the global warming issue is so controversial.

Other than scientists, viewpoints can be based on culture, economics, and convenience. Developing countries have a much different outlook than developed countries. Economics comes into play where energy is involved. Developing countries, such as Brazil, Peru, Niger, and Sudan, burn wood for fuel; developed countries burn fossil fuels. Developing countries do not see the problem of global warming as critical as a developed country because they do not feel they have contributed to it and they have more urgent issues, such as war or famine.

Commitment and sacrifice are always more difficult than convenience. To some, it is easier to continue with the status quo than to change habits and practice energy conservation. This outlook is one of the hardest to change. No matter the reason, each of us will have an effect on the future of global warming in every country all over the world.

ADVANCES IN TECHNOLOGY AND EDUCATION

Even though the theory of global warming was first introduced in the late 1800s, it was not until the mid-1900s that the theory actually started to gain notice in the scientific community. This was largely due to advances in technology and education. The first major achievement was the development of infrared spectroscopy for measuring long-wave radiation in the 1940s. Through the ability to measure infrared radiation, scientists could determine that atmospheric CO_2 was absorbing more infrared radiation (the wavelengths related to heat), which was raising the temperature of the atmosphere. Scientists were able to determine that the wavelength of infrared radiation absorbed by CO_2 was different than that absorbed by water vapor. This is significant because, if extra CO_2 was not in the atmosphere to absorb those additional wavelengths of energy, they would be lost to space instead of heating the atmosphere. Discovered by Gilbert Plass in 1955, this was like a detec-

What Skeptics of Global Warming Are Saying	
SOURCE	COMMENT/ACTIONS
Richard S. Lindzen, professor at MIT	He is willing to bet the Earth's climate will be cooler in 20 years than it is today.
Sallie L. Baliunas, astrophysicist at Harvard-Smithsonian Center for Astrophysics	She believes that global warming is a hoax.
Exxon-Funded Skeptics, United States	Since 1990, they have spent more than $19 million funding groups that promote global warming skepticism and $5.6 million to public policy organizations that publicly deny global warming and climate change exist.
Philip Stott, professor at University of London	He questions the knowledge of the IPCC.
Patrick J. Michaels, former professor at University of Virginia	He believes that global warming models are fatally flawed and, in any event, we should take no action now because new technologies will soon replace those that emit greenhouse gases.
Bjorn Lomborg, professor at University of Aarhus, Denmark	He is the author of *The Skeptical Environmentalist,* in which he argues that a statistical analysis of key global environmental indicators revealed that environmental problems were not as serious as was popularly believed.
Competitive Enterprise Institute, United States	This group wants to "dispel the myths of global warming by exposing flawed economic, scientific and risk analysis."
James Annan, British climate scientist	He says the risks of extreme climate sensitivity and catastrophic consequences have been overstated.

tive getting a fingerprint of a thief; in this case the fingerprint belonged to the CO_2 and directly linked the presence of CO_2 with rising atmospheric temperatures—or the enhanced greenhouse effect.

Until this time, many scientists argued that the oceans would absorb most of the CO_2. In the 1950s, researchers learned that CO_2 was able to stay in the atmosphere for about 10 years. The status of the oceans then came into question, as scientists began to wonder just how much the oceans could actually absorb and whether or not the CO_2 remained in the ocean or whether it could be transferred back into the atmosphere. They concluded that the oceans could only hold about one-third of the anthropogenic CO_2—much less than originally thought.

In 1958, Charles Keeling, a climate scientist, began to track atmospheric CO_2 levels in Mauna Loa, Hawaii. Measuring CO_2 concentrations ever since, this curve—known as the Keeling Curve—has become the most recognizable worldwide symbol of global warming. The curve shows a downward trend in temperature from 1940 to 1970, but starting in the 1980s, the temperature began to increase steadily. In fact, the rate of increase was so steep that it caught the attention of many scientists— this increase was not like anything that had been experienced before in human history.

By the late 1980s—nearly 100 years after Svante Arrhenius presented his ideas—global warming was finally an accepted theory and it was formally acknowledged that climate was warmer than at any period since 1880. This was a major turning point in the history of global warming. The Intergovernmental Panel on Climate Change (IPCC) was created by the United Nations Environmental Programme and the World Meteorological Organization. The IPCC is made up of more than 2,500 scientists from all over the world and many diverse fields, such as climatology, oceanography, ecology, economics, medicine, physics, geography, and others. It is important to have all these disciplines involved because global warming's reach will be so wide that it will involve many sectors that must work together to keep our world healthy. The IPCC is a prestigious group that is considered the largest peer-reviewed scientific cooperation project in history. They have released four reports on climate change to date, the most recent in 2007.

What Supporters of Global Warming Are Saying	
SOURCE	COMMENT/ACTIONS
Dave Stainforth, climate modeler at Oxford University	He says, "This is something of a hot topic, but it comes down to what you think is a small chance—even if there's just a half percent chance of destruction of society, I would class that as a very big risk."
Dr. Rajendra Pachauri, chairman of the IPCC	He says that he personally believes that the world has "already reached the level of dangerous concentrations of carbon dioxide in the atmosphere" and called for immediate and "very deep" cuts in the pollution if humanity is to "survive."
Drew Shindell, NASA Goddard Institute for Space Studies	He believes global warming will cause serious drought in some areas. "There is evidence that rainfall patterns may already be changing. If the trend continues, the consequences may be severe in only a couple of decades."
James Overland, NOAA oceanographer	Believes that by 2050, the summer sea ice off Alaska's north coast will probably shrink to half of what it was in the 1980s. This will have a profound effect on mammals dependent on the sea ice, such as polar bears, which could become extinct.
Ilsa B. Kuffner, USGS	Says oceans are becoming more acidic due to rising CO_2 in the atmosphere. This, in turn, is destroying the world's coral reefs.
Shea Penland, University of New Orleans, coastal geologist	Mr. Penland, who died in March 2008, said the rate of sea level rise has increased significantly over recent years and warns, "We're living on the verge of a coastal collapse."

(continues)

(continued)

What Supporters of Global Warming Are Saying	
SOURCE	COMMENT/ACTIONS
World Wildlife Fund	One of their top priorities is to limit global warming and reduce emissions of carbon dioxide. "If we want to have something left to protect at all, the managers of protected areas need to assess the climate change impacts and prepare their parks for the worst."
James E. Hansen, director NASA GISS	He says, "As we predicted last year, 2007 was warmer than 2006, continuing the strong warming trend of the past 30 years that has been confidently attributed to the effect of increasing human-made greenhouse gases."
Terrence Joyce, Woods Hole Physical Oceanography Department	Concerning changes to the ocean conveyor belt, he says, "It could happen in 10 years. Once it does, it can take hundreds of years to reverse." He is alarmed that Americans have yet to take the threat seriously. In a letter to the *New York Times* last April he wrote, "Recall the coldest winters in the Northeast, like those of 1936 and 1978, and then imagine recurring winters that are even colder, and you'll have an idea of what this would be like."

Today, scientists are focusing on computer modeling to try to predict the future of climate change. If scientists can accurately model the environment, they can predict future temperatures. But climate modeling is not simple and without its uncertainties. Some scientists have questioned the consistency of readings from ocean stations, satellites, and other remote stations. Climate forecasting is still in its infancy and current models still must rely on relatively sparse databases—globally collected temperatures only go back about 100 years. In some vital areas, such as the effect of cloud cover on global warming, uncertainties have grown as climatologists have attempted to model them due to their complex behaviors (short life span, rapid movement, and unpredictability).

According to Gerald North at Texas A&M University in College Station, "It's extremely hard to tell whether the models have improved in the past five years." Climate modeler Peter Stone from the Massachusetts Institute of Technology states, "The major [climate prediction] uncertainties have not been reduced at all."

Peter Stone also remarks that "We can't fully evaluate the risks we face. A lot of people won't want to do anything. I think that's unfortunate. Greenhouse warming is a threat that should be taken seriously." Possible harm could be addressed with flexible steps that "evolve as knowledge evolves. By all accounts, knowledge will be evolving for decades to come."

The following are three basic challenges in climate modeling:

(1) detecting a warming of the globe
(2) attributing that warming to rising levels of greenhouse gases
(3) projecting warming into the future

"The detection problem seems to be almost solved," says climatologist David Gutzler of the University of New Mexico in Albuquerque, referring to the IPCC's 95 percent confidence level that the Earth has warmed; its conclusions drawn partly from climate modeling.

The IPCC's report states that confidence in climate models has increased. Jerry D. Mahlman, former director of NOAA's Geophysical Fluid Dynamics Laboratory in Princeton, New Jersey, agrees. He states, "I'm quite comfortable with the confidence being expressed."

Original problems with models included the following: satellites acquiring temperature measurements of the atmosphere instead of the Earth's surface, the inability to capture daily temperature ranges, the inability to model cloud dynamics, and the inability to capture increased nighttime temperatures. Today, satellites have been calibrated to acquire temperatures from the Earth's surface. In addition, modeling climate processes, such as ocean heat transport, are more realistic.

Fudge factors that artificially steadied background climate have been eliminated and weather events such as El Niño are now more realistically represented. Improved models are also being driven by better designated climate forcings, such as variability in solar output and volcanic emissions of particulates into the atmosphere.

According to Jeffrey Kiehl of the National Center for Atmospheric Research (NCAR) in Boulder, Colorado, "We have made progress, but sometimes progress means you learn you need to know more." In particular, he refers to the pollutant hazes created by aerosol particles from wildfires and fossil fuels. He remarks, "The more we learn about aerosols, the less we know."

Another source of uncertainty is the human factor. According to Jerry Mahlman, "Social uncertainty is hard to discuss because we don't have a clue how people are going to react 30 years from now." As scientists find sensors where measurements may need to be calibrated, adjustments are made and models are rerun and verified. In an ongoing process, the scientific community, as well as the IPCC, strives to keep data integrity in their models so that their models, conclusions, and predictions are reliable, because important decisions that affect all our lives now and in the future are made from them. This is no small feat, however, because of the sheer complexity of climate systems. Because of its complex nature, it reinforces, once again, the need for international cooperation.

THE GROWTH OF ENVIRONMENTAL AWARENESS

Another factor that has helped focus attention on global warming in recent years is the growth of environmental awareness. Over the last few decades, there have been several notable anthropogenic environmental disasters. These have made the public more aware of environmental damage; sometimes permanent.

One of the worst anthropogenic disasters was the explosion at the nuclear power plant in Chernobyl, USSR, on April 26, 1986. A reactor exploded during a failed cooling system test and ignited a massive fire that burned steadily for 10 days. The accident released radioactivity 400 times more intense than that of the Hiroshima bomb in World War II. The accident affected a huge area—the plume drifted over Europe, even to North America. Ukraine, Belarus, and Russia (an area of about 90,000 square miles; 233,000 km²) were badly contaminated with unhealthy levels of radioactive elements. More than 336,000 people were evacuated. The town of Pripyat, which was built specially to house the employees of Chernobyl, was evacuated. Today, more than 20 years later, the town has never been reoccupied. The 19-mile (30-km) area around the site is known as the Zone of Alienation. All residential, civil, and business activities are prohibited by law. People who lived in the area at the time of the accident who were not killed outright suffered many serious health problems, including radiation sickness, thyroid cancer, leukemia, and birth defects causing cancer and heart disease. Researchers have estimated that roughly 7 million people were affected by this accident.

Another major anthropogenic environmental disaster was the nuclear accident on the Susquehanna River in Pennsylvania at Three Mile Island in March 1979. This was the disaster that started the controversy in the United States over the safety of using nuclear energy. In this incident, the water pumps in the cooling system failed, causing cooling water to drain away from the reactor, which partially melted the reactor core. This accident released about 1/1000 of the amount of radiation as Chernobyl did. Scientists do not know for certain how much radiation was released during the accident. The reactor core escaped meltdown just in time because of the implementation of safety measures. There was an evacuation of a five-mile (8-km) radius as a safety precaution. Experts believe that several people died of exposure to radiation. Dairy farmers reported the deaths of many of their livestock, and some local residents developed cancer. Some studies also indicate premature death and birth defects resulted as well. The cleanup for the accident began in August 1979 and finally ended in 1993; the cost was $975 million. Nearly 100 tons (91 metric tons) of radioactive fuel was removed from the area.

Times Beach, Missouri, was the site of another well-known environmental disaster that got the public's attention. High dioxin levels were found in the soil. Dioxin is a hazardous chemical used in Agent Orange, a highly toxic chemical warfare agent. Levels in the soil were determined to be 100 times higher than the threshold considered toxic

Three Mile Island was the site of one of the nation's worst anthropogenic environmental disasters. *(EPA)*

for humans. The dioxin had been mistakenly added to an oil mixture that was used to spray the roads in the 1970s to keep dust under control. Many illnesses, miscarriages, and animal deaths at the time were blamed on the levels of dioxin in the area. This disaster spurred the enactment of federal legislation, Comprehensive Environmental Response Compensation and Liability Act (CERCLA), commonly referred to as the Superfund, because it established a fund to help with the cleanup of locations of environmental disasters.

In January 2000 in the Baia Mare gold mine in Romania, cyanide used to purify gold from rocks overflowed into the Tisza River. The wastewater that contained the cyanide also contained lead and other hazardous materials. By February, it had reached the Danube, a major river that flows through or borders 10 European countries. It poisoned many fish in the river and made many people along the river ill from eating contaminated fish.

According to the U.S. Environmental Protection Agency (EPA), Love Canal is one of the most horrifying environmental tragedies in U.S. history. Love Canal was originally planned as a perfect community on the eastern edge of Niagara Falls in New York. In order to generate power for the community, a canal was to be dug between the upper and lower Niagara River so that power could be generated inexpensively for the residents. The canal was never finished. In the 1920s, all that was left was a ditch, and it was turned into a chemical dumping ground. Then, in 1953, the Hooker Chemical Company, then owner of the property and the canal, covered all the waste with dirt and sold it to the city for the sum of one dollar.

The city used the ground to build a new development and constructed about 100 homes and a school on it. During an extremely wet period, the process of leaching began. The waste disposal drums that were buried underground began to corrode and started breaking up in residents' backyards. Vegetation in the area—trees, shrubs, gardens, and grass—began to die and turn black. Chemicals began to pool in people's backyards and basements. Children would get burns on their hands and faces when they played outside. Birth defects were on the rise as well. At the time, according to a report issued by the EPA, one father whose child was born with birth defects remarked: "I heard someone from the press saying that there were *only* five cases of birth defects here. When you go back to your people at EPA, please don't use the phrase '*only* five cases.' People must realize that this is a tiny community. Five birth defect cases here is terrifying."

On August 7, 1978, New York governor Hugh Carey told the residents at Love Canal that the New York State government would purchase the homes affected by the chemicals. The same day, President Jimmy Carter approved emergency financial aid for the area. These were the first emergency funds ever to be approved for a man-made rather than a natural disaster. In addition, the U.S. Senate approved a sense of Congress amendment saying that federal aid should be forthcoming to relieve the serious environmental disaster that had occurred.

Valley of the Drums. This site was used as a disposal area for hazard-ous chemical waste, causing serious environmental pollution. Today, it is still remembered as an example of environmental irresponsibility. (EPA)

President Carter remarked of the situation: "The presence of various types of toxic substances in our environment has become increasingly widespread—one of the grimmest discoveries of the modern era."

A total of 221 families had to be relocated as a result of this disaster. Today, agencies such as the EPA, working under governing laws such as the Clean Air Act, the Clean Water Act, the Federal Environmental Pesticide Control Act, the Resource Conservation and Recovery Act, and the Toxic Substances Control Act, strive to protect the American public and the environment.

The Valley of the Drums near Louisville, Kentucky, gained national attention in 1979 and quickly became known as one of the country's worst abandoned hazardous waste sites. Over a period of 10 years in Bullitt County, industry disposed of thousands of barrels of hazardous wastes on a 23-acre site. They had been haphazardly thrown in pits, trenches, or just strewn about. The drums sat so long exposed to the out-door elements that they began to deteriorate and leak. When it rained, the barrels would fill with water, overflow, and wash the chemicals into nearby Wilson Creek, which led to the Ohio River. Chemicals found in the drums included toluene (associated with liver and kidney damage, respiratory illness, damage to developing fetuses, and death) and benzene (causes leukemia, neurological problems, and weak immune systems). This incident was another national spur to the creation of the Superfund.

In March 1989, the *Exxon Valdez* struck Bligh Reef in Prince William Sound, Alaska, unleashing the largest oil spill in U.S. history. The oil slick spread more than 3,000 square miles (7,770 km²) and onto more than 350 miles (563 km) of beaches in Prince William Sound, at that time known as one of the most pristine and beautiful natural areas in the world. The spill polluted about 1,180 miles (1,900 km) of shoreline and was devastating to wildlife in the fragile ecosystem. It killed about 250,000 sea birds, 2,800 sea otters, 250 bald eagles, and roughly two dozen killer whales.

Oil spills can be very harmful to marine birds and mammals, such as sea otters. They can also harm fish and shellfish. Oil destroys the

The tanker *Exxon Valdez* accident in Prince William Sound, Alaska, in March 1989. NOAA responders survey the oil-soaked beaches of Prince William Sound. *(NOAA)*

insulating ability of fur-bearing mammals, such as sea otters, and the water-repelling abilities of a bird's feathers, thereby exposing these creatures to the harsh elements. Many birds and animals also ingest oil when they try to clean themselves, which can poison them.

The *Exxon Valdez* spill released more than 11 million gallons (42 million liters) of oil and cost more than $3.5 billion to clean up. It received a great deal of media attention and raised the public's environmental awareness. Many people were outraged at the damage done to wildlife and their fragile ecosystems, and this single incident in particular served as a strong reminder that human behavior can have far-reaching consequences for the environment. Although these incidences were all tragedies and should never have happened, they did help raise awareness of the serious, and often tragic, effects that people's actions can have on the environment.

One topic that has been in the news recently that has made an impression on the public is the melting of the polar ice. *Time* magazine has run several special editions covering the melting of glaciers, rising seas, and diminishing icepack (April 9, 2001, April 3, 2006, April 9, 2007, and October 1, 2007, editions). *National Geographic* featured an article on melting icecaps and rising sea levels in their June 2007 issue. Environmental scientists say that because Arctic ecosystems are so sensitive and fragile, they respond quickly when they are stressed. Because of this, they provide early warning about the effects of global warming.

Scientists at the National Oceanic and Atmospheric Administration (NOAA) and the National Aeronautics and Space Administration (NASA) believe the Arctic is the first area to react to global warming. The polar regions are predicted to be affected first by global warming because warming in these regions is enhanced by positive feedback at a level unlike any other area on Earth; polar areas are extremely sensitive. The major feedback mechanism has to do with polar albedo. Snow and ice's high albedos naturally reflect 80 to 90 percent of the incoming solar radiation back into space, keeping these areas frozen. But when the polar areas begin to warm due to increased CO_2 levels in the atmosphere, the snow and ice begin to melt, greatly reducing the highly reflective surface areas. As these surfaces disappear, more of the Sun's radiation is absorbed by the underlying land or sea as heat. The resulting heat then begins to melt even more snow and ice, exposing more dark surfaces to absorb additional radiation—and a positive feedback cycle is set up. It becomes self-perpetuating, enabling the process to accelerate and feed off of itself.

Another reason why the polar areas are so sensitive is that the atmosphere in these regions is extremely dry compared to the air at lower latitudes. Because of this, less energy is used to evaporate water, which keeps the energy in the form of heat.

Further positive feedback is the resultant vulnerability to vegetation shifts in the polar regions. The boundary between forest and tundra is highly sensitive to increases in temperature. Today, global warming has caused rapid changes to take place. Over the past 50 years in northern Canada, for example, tree lines have advanced 279 feet (85 m) in elevation on warm, south-facing slopes and tree density has increased significantly (up to 65 percent) on cooler, north-facing slopes.

According to Dr. Ryan Danby of the department of biological sciences at the University of Alberta in Canada, "The mechanism of change appears to be associated with occasional years of extraordinarily high seed production—triggered by hot, dry summers—followed by successive years of warm temperatures favorable for seedling growth and survival." He warns that "Widespread changes to tree lines will have significant impacts. As tundra habitats are lost, species are forced to move upward in elevation and northward in latitude. The problem is that in mountainous areas species can only migrate so high, so they get forced into smaller and smaller areas until there is no room left for them to survive in. These results are very relevant to the current debate surrounding climate change because they provide real evidence that vegetation change will be quite considerable in response to future warming, potentially transforming tundra landscapes into open spruce woodlands."

These predictions of accelerated warming in the polar regions are also supported by climate models. According to Pal Prestrud, vice chairman of the steering committee for the Arctic Climate Impact Assessment (ACIA) report, "The projections for the future show a two to three times higher warming rate in the polar areas than for the rest of the world. That will have consequences for the physical, ecological, and human systems." Most climate models now predict the following:

- increased CO_2 concentrations will lead to a polar warming greater than the global average with more warming over land than sea and maximum warming in the winter
- a decrease in the extent and thickness of sea ice
- melting of permafrost
- a retreat of ice shelves
- migration of forests/retreat of tundra
- shifts in freeze/thaw zones to the north
- changes in polar freeze/melt cycles

According to Warwick Vincent, director of the Centre for Northern Studies at Laval University in Quebec, "Climate models indicate that the greatest changes, the most severe changes, will happen earliest in the highest northern latitudes. This will be the starting point for more substantial changes throughout the rest of the planet . . . our indicators are showing us exactly what the climate models predict. I think we're at a point where it is not stoppable but it can be slowed down. And if you think about the magnitude of effects on our society, then we really need to buy ourselves more time to get ready for some very substantial changes that are ahead." Melting ice is having a significant negative impact on the wildlife in the Arctic, such as the polar bears.

As the ice melts, it keeps the polar bears from being able to hunt for food. The 2007 film *Arctic Tale* was a documentary that illustrated the fate of Arctic ecosystems that are now facing the problem of global warming. NASA has taken satellite images of the polar ice cap and has determined that it is shrinking at a rate of 9 percent each decade. They predict that if this trend continues, there may not be any ice left in the Arctic during the summer season by the end of this century. According to the IPCC, the average annual temperature in the Arctic has increased by 1.7°F (1°C) over the last century—a rate roughly twice as fast as the global average. Over the past 100 years, winter temperatures alone have warmed 3.3°F (2°C) in the Arctic, most of it occurring in the past 30 years. Permafrost melting has been observed in Russia, Canada, Alaska, and China. In some places so much ice has thawed, it has caused the ground to subside 16 to 33 feet (5–10 m), damaging homes, roads, and other structures.

Ice in the Arctic is melting earlier each year. Scientists expect this to continue as the Earth continues to warm, which will cause problems for wildlife such as the polar bear. *(Rear Admiral Harley D. Nygren, NOAA)*

Arctic sea ice is declining 3 to 7 percent each decade. Measurements taken by submarines indicate a 40 percent reduction in ice volume across the entire Arctic Ocean basin as compared to 20 to 40 years ago. Delays in the forming of sea ice and early melt are playing havoc on physical, biological, and human systems. According to the Union of Concerned Scientists, newborn seals and walrus pups do not have sufficient time to wean adequately, causing them to die. The sea ice is also used by polar bear and caribou as a migration corridor. With weather extremes, such as heavier snowfall or freezing rain, it taxes the energy of the wildlife when they migrate and search for food, killing numerous members of the herds.

The indigenous peoples of the Arctic—such as the Yupik, Inuit, and Eskimo—also depend on the natural resources currently found in the Arctic, and their survival depends on the ecosystem remaining healthy and functioning. Global warming is already having a huge impact on them. For example, delay in the formation of sea ice has decreased the length of their hunting season, leaving them with significant food

As the Arctic ice retreats, polar bears are being stressed to extinction. Lack of ice takes away valuable hunting ground and migration corridors. *(Fotosearch)*

shortages. There has also been an increase in cloudy skies, fog, and rain during the spring and summer "drying season," when their traditional foods are air-dried for winter storage.

Melting ice also contributes to rising sea levels. Today, many of the native villages that have existed for centuries along the coasts are being flooded as the sea levels rise. Global warming is threatening the identity, culture, way of life, and very existence of cultures.

The effects of the melting of the polar ice are not just confined to the Arctic region; they will be felt worldwide. The melting of the polar ice cap is like a runaway train. As the ice melts, it actually speeds up global warming. Layers of snow and ice in the Arctic act as a type of insulating, protective blanket. By their very existence, they help keep the Arctic cool. As sunlight hits the snow and ice, it is reflected back into space because snow and ice have a high albedo (are highly reflective). Because of this, the ground does not absorb the heat from the Sun. When this protective ice layer melts it allows the Earth to absorb more sunlight and get warmer. As the ground's surface gets warmer, it melts more ice, which allows more bare ground to absorb more sunlight. The process continues: heating and melting, heating and melting.

Climate change scientists have recently announced that rising temperatures are already affecting Alaska. According to Deborah Williams, executive director of the Alaska Conservation Foundation, annual temperatures have increased 4° to 5°F (6.7° to 8.3°C) and winter temperatures have warmed 8° to 10°F (13° to 17°C)—more than any other place on Earth; more than four times the global average. Permafrost is melting, causing more than 600 families so far to lose their homes. The polar bears are also feeling the effects. According to Steven Amstrup, a polar bear specialist with the U.S. Geological Survey (USGS), "As the sea ice goes, that will direct to a very great extent what happens to polar bears." Polar bears could become extinct within the next century because they have adapted to hunting on the ice. If they try to swim, they are more likely to tire and drown.

Insects that spread disease in warmer weather have begun to migrate and damage forested areas farther north in Alaska than they ever have before. In particular, the spruce bark beetle, which breeds faster in warmer weather, has been able to reproduce at a faster rate now than it ever has in the past. It has actually sped up the breeding time by eliminating delays from one hatch to another. Today, it has destroyed almost 3.5 million acres of forests in Alaska.

As glaciers and ice sheets on land melt, the meltwater runs off the land and enters the oceans. This causes sea levels to rise; which then causes other areas to have their beaches flooded and eroded. This will affect people worldwide. Some climate change scientists have predicted the oceans could rise three feet (1 m) by the end of this century, significantly flooding many coastal areas.

PUBLIC AND MEDIA RESPONSE

Public perception is critical to the future of global warming. Whether people understand that global warming is an important issue that must be dealt with will determine the future of the Earth. Many people's perceptions are shaped by the media. Information learned at school and conversations with adults are also important sources of information for students. It is important to become educated about the topic. Be aware of what scientists know, what they suspect, and of the controversy surrounding it. There are many organizations both in governments and the private sector with Web sites that have information about global warming and the environment. These include the Environmental Defense Fund, the World Wildlife Fund, the National Wildlife Federation, Greenpeace, the Nature Conservancy, the Wilderness Society, the *National Geographic,* the IPCC, the EPA, and many others. A listing of more sites can be found in the Further Resources section.

Scientists who study global warming agree on specific key points:

- The greenhouse gases currently in the Earth's atmosphere are what keep the Earth warm and habitable. Natural levels of greenhouse gases are vital to life. Without them, the Earth would always exist in an ice age. However, too high a level of greenhouse gases is already causing harm to present-day systems, and increasing amounts would cause unknown changes to those systems.
- The concentration of greenhouse gases is increasing. Since CO_2 concentrations in the atmosphere began to be regularly mea-

sured (see the Keeling curve, discussed in chapter 1), the record indicates that CO_2 is building up in the atmosphere. Humans are increasing the concentration of CO_2 in the atmosphere, as well as other greenhouse gases, through various activities, such as agribusiness, burning fossil fuels, deforestation, and development in wetlands, and many other practices.

- Scientists theorize that if the concentration of greenhouse gases continues to rise, the Earth may become uninhabitable. This is supported by observations of other planets and the use of sophisticated computer models.

Global warming is controversial because so many issues remain theoretical. As technology improves and more research is completed, solid answers will be available and less controversy will remain. One of the biggest problems with controversy is not that people are arguing; it is that while arguing, time is being wasted and important decisions are not being made. The time to make critical decisions about changing from nonrenewable to renewable energy resources is *now*. Every day, every week, more polar ice melts, sea levels rise, flooding occurs, droughts occur, diseases spread, and intense weather causes damage. Some of the issues being argued today include:

- How much the Earth's atmosphere will heat in the future
- Whether warming is part of a natural cycle
- The cause of warming
- Global warming's effects on the oceans
- Global warming's effects on life-forms
- The explanation for the warming of the past 100 years

Action to reduce greenhouse gases needs to be taken now, before the controversy is settled. When CO_2 and other greenhouse gases enter the atmosphere, they stay for many years. There is no time to waste.

The Big Picture

This chapter focuses on looking at warming as a global issue and a global system. It will present what climate scientists refer to as climate forcing—the positive and negative feedbacks of the climate system and what they mean to the issue as a whole. Next, it will introduce the concepts of scientific inquiry and the limits of technology, their common ground and the boundaries between them. The structure, goals, and purpose of the Intergovernmental Panel on Climate Change (IPCC) will be addressed, as well as what scientists currently know, speculate, and do not yet know about global warming. Finally, the chapter will glance at what lies ahead.

CLIMATE FORCING—POSITIVE AND NEGATIVE FEEDBACK

The Earth's heat budget is controlled by the amount of energy it receives from the Sun and subsequently returns to space. The Earth's global energy balance is maintained when the incoming energy from the Sun is balanced with the outgoing heat from the Earth. There are mechanisms that cause the Earth's energy budget to get out of balance. Then it can result in climate change.

There are several mechanisms that can change the Earth's energy balance in either a positive or negative way. Some of these mechanisms are natural and some are human-caused. One of the natural mechanisms that can affect the Earth's energy balance, and therefore climate, is fluctuations in the Earth's orbit. This includes the properties that were discussed in chapter 5: changes in the Earth's eccentricity, tilt, and precession. As the orbit becomes more eccentric, tilt becomes more pronounced, or precession shifts, the Earth's climate experiences changes. Changes in ocean circulation can also affect climate by moderating temperatures of ocean water and air masses transported around the globe. Volcanic eruptions are another natural mechanism that can have an effect on the energy balance. When a significant amount of particulates are added to the atmosphere from a powerful volcanic eruption, they can effectively block out incoming radiation and lower temperatures on Earth, cooling climate for a certain period of time. There is also a natural variability in solar radiation as sunlight goes through its natural cycles of energy fluctuation. The balance can be altered by reflection from clouds or gases, absorption by various gases or surfaces, and emission of heat by certain materials. Variations in the amount of active gases present also can change the energy balance.

However, in the past several decades, changes in the composition of the Earth's atmosphere that have seriously affected the energy balance have not been natural: they have been caused by human actions. In particular, they have been in the form of air pollution due to the emission of greenhouse gases. They end up forcing the climate to change. Scientists refer to these various natural and human-induced mechanisms as climate forcing mechanisms.

Scientists refer to climate forcing in both the positive and negative sense. According to Gavin A. Schmidt of the National Aeronautics and Space Administration (NASA), both the positive and negative forcings must be taken into account in order to build a reliable model to base future predictions on. A positive forcing is a forcing that warms

Excessive amounts of ash from volcanic eruptions can cause cooling because the ash and other particulates block out incoming solar
radiation. This photo shows the eruption of Mount Pinatubo on June
12, 1991. It caused a noticeable worldwide cooling. *(Dave Harlow,
USGS)*

the system. A negative forcing cools the system. When radiative forcing (change between the incoming radiation energy and the outgoing
radiation energy in a climate system) occurs, it alters the equilibrium
and causes a new balance to be reached and maintained.

It is important to understand radiative forcing because as these
factors exert an influence on the Earth's energy balance, they directly
affect climate and climate change. Scientists who study global warming
pay specific attention to these various forcings when they build mathematical models in an attempt to better understand climate behavior.
As shown in the figure on page 150 there are distinct forcings important to the study of global warming. Those elements that contribute
a positive forcing (and increase global warming) include greenhouse
gases, stratospheric water vapor, and black carbon on snow. Negative
forcings—elements that counteract global warming—include stratospheric ozone and the albedo associated with certain types of land use.

The figure makes clear that the net anthropogenic component is quite significant, supporting the theory that it is largely human activity causing global warming.

There is nothing we can do about natural forcings, i.e., when the Sun's intensity changes. Humans, however, do have control over forcings caused by emissions of greenhouse gases from the burning of fossil fuels and methane released by production of animals for food, reduction in CO_2 storage through deforestation, adding CO_2 to the atmosphere through the burning of forests, and deliberate changes in land use that affect global warming. If humans choose wisely and curtail these detrimental activities, they can control much of the radiative forcings that currently act on the environment to accelerate global warming. When scientists talk about changes in global warming and the damage done to specific ecosystems, they often put it into context with climate forcing values because these tell scientists much about what has happened in an

Industrial pollution is a human-caused positive forcing that serves to increase the atmospheric temperature by adding greenhouse gases. *(Nature's Images)*

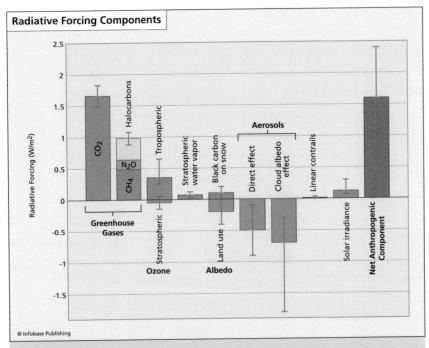

This graphic illustrates the concept of radiative forcing. In order to understand global warming, it is necessary to understand the components in the atmosphere that control the overall warming and cooling on a short- and long-term basis. *(modeled after NASA)*

area, what will likely happen, and where focus needs to be directed to offset unnecessary negative impacts.

SCIENTIFIC INQUIRY AND THE LIMITS OF TECHNOLOGY

Scientists are always asking and answering questions. In order to get answers, they must follow steps through discovery and analysis. Formulating a hypothesis and designing an experiment to test the hypothesis are the first steps in all scientific inquiry.

Once a test has been designed, data must be collected. When collecting data, it is important to maintain data integrity and obtain a representative sample of what is being measured. Then data needs to be interpreted or analyzed. The results of the analysis enable the scientist to draw final conclusions. Depending on the result, the scientist can adjust his hypothesis and run more experimentation or take the study further by posing new questions that may, or may not, relate to the original question. This can lead to new investigations and discoveries.

Scientific inquiry requires keen, discriminating observance. It also requires the ability to group related ideas and objects and interpret and quantify data and explain its significance. It also requires deductive reasoning to reach conclusions.

Through scientific research, scientists have learned that nature has a predetermined order, and they seek to find and characterize its patterns. As each new piece of scientific knowledge is gained, science is advanced. Advancements build upon previous advancements. Because these advancements were made with carefully tested procedures, the results are also repeatable, making them solid enough to develop theories.

Because of the rapid spread of technology, some people believe that answers to all scientific questions will eventually be obtainable. This may not be the case. There are some very real limits to technology and what is possible. Some goals or achievements may never be possible. It is important to understand the limits of technology as it relates to science. This is especially relevant to global warming. As changes are needed to control and adapt to the problem, it is important to know what is technically feasible and what is not and to work within that framework.

Technology is limited by adherence to physical laws. It must follow specific, predefined steps that may not be technologically possible. For example, nuclear fusion is understood in the scientific community, but limited by technology. Once the technology is developed, it is believed it will be the new cheap, safe, clean, and almost limitless energy source of the future.

There are also limits imposed by equipment, such as computer hardware and software. Sometimes it might cost too much to develop software to handle a specific problem or developing the required equipment may not be feasible. There are also mathematical constraints. In order to be able to create some of the models necessary to mimic weather and climate, incredibly complicated mathematical algorithms must be written. The Earth's climate system is so complicated, with so many interactive variables on both long- and short-term scales, that the knowledge and ability of scientists is continually challenged when trying to create realistic, reliable, and accurate models. As scientists continue to learn and understand more about this complicated system, some of these mathematical constraints are eliminated; but mathematical constraints do present a critical challenge.

There are also design and developmental limitations in creating climate models to measure, analyze, and assess global warming. Again, because the Earth's climate is such a complicated system, with so many different subsystems, writing software code for models is an enormous task. In addition, there are several parameters that can be assessed in the global warming problem—both natural and anthropogenic, such as wind speed, wind direction, temperature, pressure, circulation, clouds, presence of continents, presence of ice, etc. The ways these work together and separately can also contribute to different outcomes. Models must take all of these contingencies into consideration, as well as unexpected or unexplained scenarios.

Economic limitations also crop up. Scientific research and advancements are costly. It takes significant investments of capital to fund scientific research, to buy equipment to run labs, and to invest in hardware/software development. Because of this, many countries cannot afford to invest in and carry on research toward understanding and finding a solution to global warming. Often, in developed countries, it is the federal, state, and local government agencies that come up with the funding for scientific research and development. That is why much of the data on global warming is housed at agencies like the National Oceanic and Atmospheric Administration (NOAA), EPA, NASA, and U.S. Geological Survey (USGS). Research and work accomplished by universities is also often funded by government agencies.

THE INTERGOVERNMENTAL PANEL ON CLIMATE CHANGE (IPCC)

The Intergovernmental Panel on Climate Change (IPCC) is a group of approximately 2,500 scientists working together to study the problem of global warming and focusing on a workable strategy to deal with the problem before it is too late. The IPCC was formed jointly in 1988 by the United Nations Environment Programme (UNEP) and the World Meteorological Organization (WMO) because of a growing worldwide concern over global warming. It is considered to be the most technical and authoritative group of experts on global warming in the world. The scientists who make up the IPCC are the world's top scientists in all relevant fields, such as physics, chemistry, geography, geology, medicine, economics, sociology, botany, biology, and others. Together, they write and review scientific literature on global warming issues and produce authoritative findings on the status of global warming and its effects on the Earth and its climate.

The IPCC has been studying global warming for many years and has released four reports over the past 17 years: in 1990, 1996, 2001, and 2007. In the 2007 report, Dr. Rajendra Pachauri, the chairman of IPCC, reported at an international conference in Mauritius, which representatives of 114 governments attended, that he personally believes that the world has "already reached the level of dangerous concentrations of carbon dioxide in the atmosphere" and called for immediate and "very deep cuts" in the pollution if humanity is to "survive." He also said that "climate change is for real. We have just a small window of opportunity and it is closing rather rapidly. There is not a moment to lose." He based his remarks on the following observations:

- Coral reefs throughout the world are perishing as the oceans warm up. So far, up to a quarter of the world's corals have been destroyed.
- Based on a multiyear study by 300 scientists, the Arctic is warming twice as fast as the rest of the world. The ice has melted so fast, it has shrunk 20 percent in the last 30 years alone.
- Arctic ice is also 40 percent thinner today than it was in the 1970s and is expected to be completely gone by 2070.
- The effects of global warming we are feeling today are due to the CO_2 put in the atmosphere in the 1960s. When we feel the greater effects of increased pollution from later decades it will be even more severe.

The IPCC issued its fourth report in 2007, six years after its third report. Because scientists like to project their findings in terms of probabilities,

One of the largest contributors to greenhouse gases through the burning of fossil fuels is automobiles. The United States is the world's largest consumer of petroleum. *(Nature's Images)*

leaving an opening for the possibility of new findings, the IPCC report is also given in terms of probabilities. But an interesting thing happened in the 2007 report that had not happened before. The IPCC is made up of many scientists from many diverse fields, countries, and backgrounds. In the past the IPCC's reports have been very conservative. The 2007 report, however, has a much different tone. This report was worded much more forcefully than ever before.

This time the IPCC said the likelihood was 90 to 99 percent (labeled "very likely") that emissions of heat-trapping greenhouse gases like CO_2, originating from smokestacks and automobile tailpipes, were the dominant cause of the observed warming of the past 50 years.

This is a significant change. In its first report (1990), the IPCC found evidence that global warming was real but concluded it could be caused as much by natural factors as human factors. Then, in its 1996 report, it changed its stand and supported the idea that most of the evidence suggested a definite human influence on global climate. In its third assessment (2001), it raised the human component and said that the probability of human activity being responsible for most of the warming of the 1900s was likely—the equivalent of a 66 to 90 percent probability. Finally, in its most recent report (2007), it has classified human activity as 90 to 99 percent (very likely) the major force behind global warming. When that many scientists are working together, giving an overwhelmingly high probability consensus on such a controversial issue, it stands as a strong testament to the severity of the threat of global warming. Scientists have shown that doubt and skepticism have given way to certainty. Overwhelmingly, the world's foremost authorities on global warming agree that the issue has become a risk to the Earth and humanity.

It is important to note that even though the general consensus supports the realities of global warming, scientists have not closed their minds to the issue. An integral part of scientific inquiry is that new evidence can always be revealed and either strengthen or challenge conclusions. In this regard, scientists keep an open mind that new evidence could be found to either lessen the severity of human actions on the environment or show that the situation may actually be worse than initially expected.

As noted in the *New York Times* by William K. Stevens on February 6, 2007, one significant factor is that the IPCC's conclusions have become significantly, consistently stronger over the past 17 years, even though the IPCC has lost some scientists and gained new ones throughout this time. As scientific inquiry uncovers increasing evidence of the human factor involved in global warming, the scientific consensus around the world increasingly supports the findings of the IPCC. The IPCC announced that global warming is a certainty and pointed out that 11 of the last 12 years were among the 12 warmest ever recorded. Adding to its growing list of evidence, the IPCC also noted the following in its 2007 report:

- Droughts in some areas of the world have increased in intensity and length, causing severe impacts on human settlements.
- Precipitation patterns have altered the tropical and subtropical regions (the world's principal rain forest areas). There has been

a measurable decrease in precipitation here, while other areas in the world have had measurable increases.

- Changes in precipitation intensity have occurred. Some areas are having more frequent severe storms. Several areas are experiencing less, but more severe, storm events. This can cause flooding and diminish long-term water supplies.
- The world's temperate zones are experiencing warmer average temperatures. Some areas are experiencing less cold days and frost periods and an increase in heat waves. These conditions stress resources and can also negatively affect people's health.
- The Northern Hemisphere—where most of the Earth's continents are located—is projected to warm more than the projected average.

The IPCC has attempted to increase the awareness of and concern for global warming in all countries of the world. The participating scientists

One effect of global warming is drought. As global temperatures increase, some areas will experience shortages of water, which will have an adverse impact on their ability to grow food, maintain proper hygiene, and maintain good health. This is an especially serious threat to developing countries. *(Nature's Images)*

Extreme weather events are another result of global warming, such as these multiple cloud-to-ground and cloud-to-cloud lightning strikes during a nighttime electrical storm. *(C. Clark, NOAA)*

continue to conduct research to further scientific knowledge to enable countries to take the appropriate steps to deal with the problem.

WHAT SCIENTISTS KNOW, SPECULATE, AND DO NOT KNOW ABOUT GLOBAL WARMING

Because the Earth's climate is a huge, complicated system of many components (such as atmospheric temperature, humidity, wind systems, ocean currents, and the influence of topography) at many different scales (local, regional, and global), there are many things scientists do know about global warming, but there are also some things they still do not know.

According to the EPA there are three things scientists do know about global warming.

1. They know for certain that human activities are responsible for changing the composition of the Earth's atmosphere. The increas-

ing levels of CO_2 and other greenhouse gases in the atmosphere since the preindustrial period have been well documented. There is nothing other than human activities that can explain the addition of these additional greenhouse gases.

2. Scientists clearly understand the greenhouse effect. They know that the Earth's natural greenhouse effect is vital for life to be possible on Earth. They also understand that human activities are adding additional CO_2 that enhances the natural effect and raises temperatures. Through experiments, they also know that the CO_2 that humans add can stay in the atmosphere for centuries.

3. Temperature records kept since the late 1800s have enabled scientists to determine that the atmosphere has warmed an average of 1°F (0.6°C) over the past 100 years worldwide, except in the polar regions where it has experienced an even greater rise. This is supported by much evidence: glaciers are melting, snow cover in the Northern Hemisphere is melting, and regions with permafrost are starting to thaw.

Sometimes scientists classify things as "likely but not certain." This leaves the door open for further evidence to be collected to either lend credence to the theory or not. Based on the EPA's findings, scientists are not certain how much warming can be attributed to human interference and how much has been contributed by natural factors because the interactions of multiple variables, both anthropogenic and natural, are so complex. It is like untangling a strand from a giant web. Research in this area is incomplete.

It is uncertain how much and how fast the Earth's atmosphere will warm. The IPCC has projected a 2.2–10°F (1.4–5.8°C) increase by the year 2100, but they cannot say with any certainty how much it will warm by a specific time period.

There are also things scientists admit they do not know about global warming. Scientists have not been able to predict, for example, the precise effects global warming will have on individual communities—they have only been able to describe regional effects with accuracy. Predicting changes in temperature, rainfall, water supply, and soil moisture for local areas is not possible at this time. They also have not been able to predict specific impacts on local health, forests, wildlife, resources, agriculture, and water supplies due to global warming. One of the reasons this is so difficult to predict is because computer models have not progressed far enough to analyze small-scale local effects; they work better at large-scale global effects.

Another source of scientific uncertainty stems from the fact that complex systems inherently have unexpected surprises associated with them and are usually hard to predict and account for ahead of time. Major uncertainties include:

- How much more warming will there be?
- How fast will the warming happen?
- What will be the adverse effects?
- Will there be any beneficial effects?

Scientists have learned to deal with, and accept, uncertainty. Uncertainty leads to more research, which ultimately leads to discoveries and better understanding, offering a new framework to operate in and advance.

WHAT LIES AHEAD

Because the warmest recorded years on record have all occurred recently and scientists worldwide acknowledge that levels of atmospheric CO_2 are continually rising, it has become evident that the problem of global warming is not just going to go away. It is going to take the involvement and commitment of the scientific community to study the issue and search for feasible solutions. Governments worldwide will need to work together to help the environment. Every person on Earth will have to do their share in helping create a cleaner, healthier environment.

Climate change scientists have determined that the poorer nations will be more vulnerable to the negative effects of global warming. It is the wealthier nations that burn large amounts of fossil fuels that are adding the most to the problem, but as global warming leads to extreme weather disasters, droughts, floods, heat waves, food shortages, sea level rise, and the spread of disease, it will be the poorer nations that suffer the brunt of the hardships because of a lack of resources.

Opinion surveys have been taken in America, such as a Gallup Poll in 2003, which found three out of every four people in the United States would support mandatory controls on CO_2 levels. This is significant, because it implies that most Americans would like to see their government take positive action on global warming. The United States must implement policies in line with other countries to combat global warming. For example, people in western Europe rely heavily on public transportation to commute each day, European restaurants typically do not provide ice water with dinner in order to conserve water, recycling programs are heavily supported, composting is seen in backyards, and many homes use wind and solar energy. Many countries, such as Sweden, Norway, Great Britain, and Canada support productive public education programs, keeping their citizens informed about the health and condition of the natural environment, as well as ways they can help toward solving the global warming problem now and in the future.

Although not widespread at this point, some of these same types of actions are occurring in the United States under the direction of individual states. California has taken the lead in combating global warming by fighting pollution, supporting renewable energy, and cutting back the use of fossil fuels.

Conclusions and a Glance into the Future

This chapter looks at how to focus on global warming in relation to the scientific consensus. It examines the major contributors of global warming. It offers some wisdom from experts on warming, and, finally, it focuses on coping and adapting to changing climates and how research and learning can keep us moving ahead in the right direction.

AN OVERWHELMING CONSENSUS

The majority of the scientific community is no longer debating the basic facts of climate change. They have been accepted by the IPCC and backed by the U.S. National Academy of Sciences and the American Geophysical Union. According to Gabriele Hegerl, associate research professor at Duke's Nicholas School of the Environment and Earth Sciences, who was a coordinating lead author of the IPCC report's chapter on "Understanding and Attributing Climate Change," the IPCC's

2007 assessment "gives a very balanced view of the evidence for climate change, predictions of future change and the remaining uncertainties, and it draws input from a very large number of scientists worldwide. The information in the report will be very important to develop effective policies to address global climate change and to prepare for the change that is coming our way."

To see the signs of change, one only has to look. Climate change experts from the National Oceanic and Atmospheric Administration (NOAA) report that in the Northern Hemisphere, spring is arriving sooner and fall later. NOAA has also determined that sea level has already risen worldwide by about 6 to 8 inches (15–20 cm) during the last century. Roughly 1 to 2 inches (2.5–5 cm) of the rise has resulted from the melting of mountain glaciers. One example of melting glaciers can be found in Glacier National Park in Montana. The National Park Service reports that the glaciers there are now approximately one-third the size they were in 1850. To support the theory of global warming, climate reports at the park dating back to 1900 show a correlation between the rate of glacier retreat and the regional climate.

Another 0.7 to 2.8 inches (1.8–7 cm) of sea-level rise has resulted from the expansion of ocean water that comes from warmer ocean temperatures as global warming heats them up. When experts study global warming issues through the development of climate models they predict the Earth will warm 1.8 to 6.3° F (1–3.8°C) by the year 2100, with a best estimate of 3.6°F (2.2°C). According to the experts, the time to act is now.

Interestingly, even groups that have traditionally been skeptics of global warming, such as the Competitive Enterprise Institute, backed down in April 2007 and acknowledged that global warming and its human influence was indeed real. Over the past several years, many who had been outspoken in their denial of global warming have changed their opinions as the evidence of human-induced global warming has continued to mount.

Another issue concerns the Kyoto Protocol. The Kyoto Protocol (the Protocol) is an amendment to the United Nations Framework Convention on Climate Change. Its objective is the reduction of greenhouse gases that contribute to global warming. It was adopted on December 11,

1997, at the Third Conference of the Parties to the treaty when they met in Kyoto, Japan, and entered into force on February 16, 2005. As of April 2008, 178 parties have signed and ratified the Kyoto Protocol and committed to reduce greenhouse gas emissions to the levels specified by the treaty.

The United States has not ratified the Kyoto Protocol, however. Although the Clinton administration did sign the Protocol in 1998, on March 29, 2001, the Bush administration withdrew the United States from it because they believed the Protocol was "fundamentally flawed and was not the correct vehicle with which to produce real environmental solutions." The following statement from Washington, D.C., was issued on March 29, 2001: "The Kyoto Protocol does not provide the long-term solution the world seeks to the problem of global warming. The goals of the Kyoto Protocol were established not by science, but by political negotiation, and are therefore arbitrary and ineffective in nature. In addition, many countries of the world are completely exempted from the Protocol, such as China and India, who are two of the top five emitters of greenhouse gases in the world. Further, the Protocol could have potentially significant repercussions for the global economy."

WHO ARE THE MAJOR CONTRIBUTORS?

Countries do not emit equal amounts of CO_2. Although there are several ways that CO_2 can enter the atmosphere, such as through the burning of wood for fuel and deforestation, most of the CO_2 is added through the burning of fossil fuels. The most common activities that burn fossil fuels for energy are industrial processes, energy production, and transportation. Thus, it is the developed countries that contribute mainly to global warming.

Another major source of CO_2 is from changes in land use. The biggest impact is seen where forested areas are cut down and replaced by grasslands to graze cattle, farms for agriculture, or urban areas. Many of these areas include the world's rain forests, which store immense amounts of CO_2 in their biomass. While the CO_2 is stored in the trees' trunk and leaves, it is kept out of the atmosphere. If the forest is burned down, it not only releases all the stored CO_2 to the atmosphere, but it

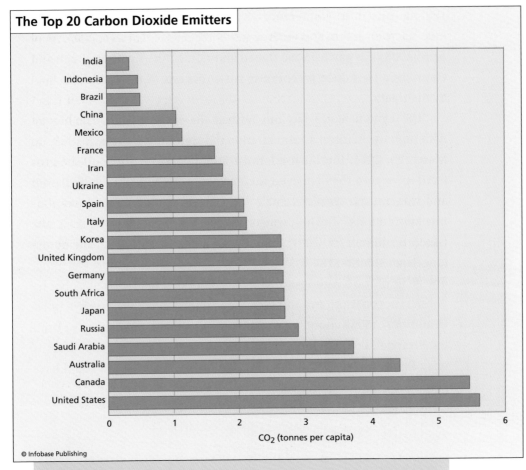

The Top 20 Carbon Dioxide Emitters

CO$_2$ (tonnes per capita)

© Infobase Publishing

This graph shows the world's main CO$_2$ polluters. These values reflect the world carbon dioxide emissions from the use of fossil fuels as of 2004 (the most recent available). *(U.S. Department of Energy)*

takes away the natural carbon storage reservoir that the trees actually were, making it so that future CO$_2$ will not be able to be stored there. This problem is significant in Africa, Asia, and South America.

One of the biggest consumers of fossil fuels for energy and emitters of CO$_2$ is industrial farming. Conventional food production and distribution require a huge amount of energy. Roughly 10 percent of the energy used annually in the United States is by the food industry. According to a study conducted in 2002 by Johns Hopkins Bloomberg School of Public Health, three calories of energy were needed to create one calorie of edible food. Some foods, however, take even more, such as grain-fed beef, which requires 35 calories for every calorie of beef produced. Even worse, that estimate does not take into account the energy used in processing and transporting the food. With those two variables taken into account, another seven to 10 calories of input energy to produce just one calorie of food must be added. The biggest user of fossil fuel energy in industrial farming is for chemicals. Up to 40 percent of the energy used in the food system goes toward the production of artificial fertilizers and pesticides, which are derived from atmospheric nitrogen and natural gas. Their production and distribution takes an average of 5.5 gallons (20.8 l) of fossil fuels per acre. The transportation of food is another major user of fossil fuels. The sectors that emit the largest amounts of CO_2 include the following:

- Power stations (21.3 percent)
- Industrial processes (16.8 percent)
- Transportation fuels (14 percent)
- Agricultural byproducts (12.5 percent)
- Fossil fuel retrieval, processing, and distribution (11.3 percent)
- Residential, commercial, and other sources (10.3 percent)
- Land use and biomass burning (10 percent)
- Waste disposal and treatment (3.4 percent)

This includes activities such as the burning of fossil fuels, deforestation, land use change, livestock manure management, paddy rice farming, wetland changes, pipeline losses, covered vented landfill emissions which leach methane, vented septic systems, the use of chlorofluorocarbons (CFCs) in refrigeration systems and halons in fire suppression systems and manufacturing processes, and agricultural activities such as the use of fertilizers that lead to higher nitrous oxide concentrations.

In October 2008, scientists from the Scripps Institution of Oceanography released disturbing news. The compound, nitrogen trifluoride (NF3), used in the manufacture of flat panel TVs, computer displays, microcircuits, and solar panels is 17,000 times more potent than CO_2 as a greenhouse gas. Not only that, but new analytical techniques have allowed scientists to measure the amount of this gas in the atmosphere. (It had been estimated to have been less than 1,200 metric tons in 2006.)

The new technology shows that in fact it was 4,200 metric tons in 2006. In 2008, there was around 5,400 metric tons in the atmosphere, an amount increasing at close to 11 percent a year.

Another major concern with global warming is the massive release of methane. A greenhouse gas 20 times more potent than CO_2, it is being released into the atmosphere from beneath the Arctic seabed. As the Arctic becomes warmer, massive amounts of methane are bubbling to the surface. This has been documented in the East Siberian Sea. Methane is also being released from thawing permafrost and frozen tundra in Siberia. This mechanism is the same one that scientists believe may have ended past ice ages. Recent discoveries have also revealed the existence of a vast band of methane hydrate ice along the world's continental slopes at 1,640 feet (500 m) depth.

There has been a drastic increase in CO_2 emissions since 2000 by China, which is responsible for most of the global growth in emissions from 2000 to the present. Emissions from Russia have actually decreased due to more efficient energy use, and CO_2 emissions from industry in Europe have been roughly stabilized since 1994.

The peak in Asia that China has recently set, however, is a sign of the beginning of industrialization in China during their progress toward becoming a developed country. Over the time period of 2000 to 2010, it is predicted that China will continue to increase its CO_2 emissions due to rapid construction of old-fashioned power plants. According to the Netherlands Environmental Assessment Agency, the largest producer of CO_2 emissions since 2006 has been China with an estimated annual production of 6,200 megatons, followed by the United States with 5,800 megatons. Another country rapidly rising in significance as an emitter of CO_2 is India. They are currently the fourth largest emitter. As they continue to industrialize and develop, they will continue to add significantly to the problem.

Many developing countries are trying to elevate their standards of living today by becoming more industrialized. This presents a problem with the global warming issue; especially if developed countries cut back but developing countries do not. Unfortunately, right now there are no international agreements in place to handle this inequity. While the Kyoto Protocol deals with the issue of greenhouse gases, one of its shortcomings is that it lets China and India off the hook of being held accountable for their CO_2 emissions during modernization.

WHAT THE EXPERTS SAY

There is a great amount of scientific research going on worldwide to further understanding of global warming—studies conducted by governments, universities, research organizations, and conservation groups. The knowledge is widely shared because this is a global problem, and the effort to fix it must be global also.

The Earth Is Heating Up

In February 2005, Dr. James Hansen, head of the NASA Goddard Institute for Space Studies (GISS), analyzed the temperature data compiled for 2004 and determined that the Earth's surface temperature worldwide was 0.86°F (0.48°C) above the average temperature for the time period from 1951 to 1980.

Dr. Hansen noted that, "There has been a strong warming trend over the past 30 years, a trend that has shown to be due primarily to increasing greenhouse gases in the atmosphere." Even though natural climate change does have a noticeable effect on the climate—such as El Niño and volcanic eruptions—scientists believe anthropogenic pollution plays a much more significant role in warming.

Present evidence makes scientists like Hansen interested in predicting future trends of climate warming. Hansen and his researchers have been able to figure out that the Earth's surface absorbs more of the Sun's energy than what gets reflected back to space. Hansen and his team have also determined that global warming has reached a serious enough level that it has started to affect the seasons, making them warmer than they traditionally have been.

The Earth's Energy Balance

Climate scientists study the Earth's energy balance when they try to determine how much the Earth is warming up/cooling down, where it is, when it is, and how different regions affect each other. In April 2005, the GISS completed a study of the Earth's energy balance using several different scientific tools. They used global climate models, satellite observations, and measurements taken from ground observation at different sites. What they found was surprising. They were able to determine that the Earth's traditional energy balance is off. Scientists uncovered two alarming facts: (1) the Earth is absorbing about 0.85 watts of energy per square meter more than it is radiating back to space; and (2) a significant amount of that excess energy is residing in the oceans, which means humans have not felt the full negative impact yet of the energy stored in the ocean basins.

Although these amounts might not sound like much, scientists are very concerned because this energy imbalance is large when compared to the Earth's past. They have determined that an imbalance, for example, of one watt per square meter maintained more than 10,000 years is enough to melt ice equivalent to 0.6 miles of sea level.

Scientists also determined from this study that an extra 1°F (0.6°C) increase in the global temperature is bound to happen no matter what we do—even cutting back on adding CO_2 to the atmosphere. They believe that the Earth has not increased enough in temperature since 1880 to explain the total energy imbalance. Even though some of the excess heat has gone to melt snow and ice cover and warm the ground, a huge amount has remained dormant in the oceans. Scientists believe that when this stored CO_2 enters the picture, there will be an additional warming of 1 to 1.2°F (0.6–0.7°C). This lag time worries scientists. Some have expressed concern that people will not change their energy choices until they see the worst of the global warming effects, and by that time so much damage will have been done to the environment it will be irreparable.

Global Warming and Future Droughts

The scientists at GISS completed research in February 2007 linking future global warming with droughts in certain parts of the world, including the southwestern United States. They used records of the Sun's output in a model to illustrate how a climate dominated by greenhouse gases would ultimately change rainfall patterns. What they found was that the same areas that experienced droughts in ancient times would experience them again.

One of the consequences of global warming is drought. This region in Utah has already experienced drought conditions, leaving the reservoir at lower than normal capacities for the past two years, as seen by the exposed shoreline several feet above the present waterline. *(Nature's Images)*

Drew Shindell, GISS team leader, said that there is already evidence that some rainfall patterns may be changing. Examples can be seen in North Africa, the Middle East, and the Mediterranean. If these trends continue, in a couple of decades there could be serious water resource challenges for many people in the world. Their model showed water stress could occur in the southwest United States, Mexico, parts of North Africa, Australia, and the Middle East. Areas suffering drought are also more likely to have erosion problems, making it difficult to support populations. Researchers believe the same climate change and drought scenarios may have been the demise of past civilizations, such as the Pueblo people of Arizona and New Mexico who abandoned cities in the 1300s. These people had built massive tribal houses, many into cliffs, that could house 500 people or more. These villages thrived as the inhabitants traded with others as far away as South America. Then, suddenly, these civilizations disappeared, leaving priceless artifacts behind.

The Sun and Volcanoes—Their Roles in Regional Climate

Scientists are also interested in global warming's impact on smaller regional and local levels. Because of readily available data, climatologists often refer to data collected from 1600 through 1860 as a baseline to compare against current data. Human activity has drastically changed the composition of the atmosphere since 1600 and by comparing present data to the baseline data, scientists can determine how much it has changed. Before the Industrial Revolution, the biggest controller of climate change was solar variation and volcanic eruptions. Through modeling techniques, GISS scientists were able to determine that changes in incoming solar radiation have had a larger impact on regional climate than volcanic eruptions have. Solar variability cycles can last for many years, while the particles from volcanic eruptions that stay in the atmosphere and block incoming solar radiation typically affect regional climate for only a couple of years. By studying these older climate records, scientists can understand past conditions.

Antarctic Climate Change

Scientists at NASA ran a computer model in 2004 that indicated that the South Pole would enter a period of warming for the next 50 years. Because the use of ozone-depleting chemicals has been banned for several years, the hole in the ozone layer is repairing itself as ozone levels return to normal. This, in turn, will cause dominant airflow patterns in the atmosphere that weaken the westerly winds and warm the air temperature.

The model was run several times comparing the oceans to the atmosphere. Each time the model began with the year 1945 and ran through the year 2055. The results were encouraging—the results for 1945 to the present matched the actual records very well. The values that were entered to reflect greenhouse gases were taken from a combination of 1999 data and the IPCC's mid-range estimates of future emissions.

The results of the model were that the most significant impacts in Antarctica would be the melting of the ice sheets and their subsequent sliding into the ocean, making sea levels rise. NASA scientists reported thawed ice sheets the size of the state of Rhode Island have already collapsed into the ocean during warming—partially brought on by stronger westerly winds in the area that heat the peninsulas as they blow over them.

COPING WITH GLOBAL CLIMATE CHANGE— ADAPTATION

Enough research has been conducted that scientists have determined that, if CO_2 concentrations in the atmosphere reach twice their pre-industrial levels (278 ppm), global climate will probably warm 3.5 to 8°F (2.1–4.8°C). Scientists also warn that there is a 10 percent chance warming could be even higher and doubling of CO_2 concentrations could occur sometime around 2050 if immediate action is not taken to find renewable sources of energy.

Climate change scientists also warn that even if CO_2 levels increase only moderately, it will significantly alter ecosystems, water supplies, and food sources. Warming in the United States is anticipated to be about one-third greater than the global average. Americans will experience rising sea levels, enhanced water cycles resulting in more precipitation, temperature extremes, and increased storm activity. This will affect not only humans, but ecosystems as well.

There are two sides to the global warming issue: those who see it as a real threat and those who do not. There has emerged a middle group. According to Mike Hulme of the Tyndall Center for Climate Change Research in Britain, this new middle-of-the-road group believes it is the uncertainty of the global warming issue that makes it critical to act now, before it gets any worse. He believes the best strategy is to "raise public appreciation of the unprecedented scale and nature of the challenge." He also emphasizes that "climate change is not a problem waiting for a solution, but a powerful idea that will transform the way we develop." Hulme also stresses that although it is important for the public to take proactive steps, people should not panic. Panicking will help no one; it will do nothing but slow the corrective process.

Jerry D. Mahlman, a climatologist at the National Center for Atmospheric Research, in Boulder, Colorado, says, "This is a mega-ethical challenge." He also said that "the buildup of carbon dioxide and other greenhouse gases cannot be quickly reversed with existing technologies. Even if every engine on Earth were shut down today, there would be no measurable impact on the warming rate for many years, given the buildup of heat already banked in the seas." He believes that because of the scale and time lag, a better strategy is to treat human-caused warming more as a risk to be reduced than a problem to be solved.

Humans must also learn to cope with global warming. They must develop strategies to manage what is to come in the next 100 or more years. In order to do this, it is important to understand how natural systems will or can adapt. Humans will be required to adapt to the changes.

With the amount of CO_2 already in the atmosphere and stored in the oceans, warming is inevitable. The resulting failure or success of the adaptation will, in the end, be a result of the individual and collective attitudes of people. Working together and cutting back is possible. Some scientists believe that educating the public about these issues is the best approach; once the public understands what is at stake and realizes it can be part of the solution, there will be more willingness to help and adapt. There are many opportunities for Americans to cut back. The United States is home to 5 percent of the world's population, but contributes 25 percent of the greenhouse gas emissions, more than any other country in the world.

RESEARCH AND LEARNING—LOOKING AHEAD

Much research still needs to be done on global warming. Scientists have targeted specific areas where further research is needed. One area is the study of atmospheric composition. The composition of the global atmosphere has a direct influence on climate and air quality, which also affects the health of ecosystems and humans. Further research is necessary to see how human activities, as well as natural phenomena, affect atmospheric composition and how those changes relate to climate change. This type of information is necessary for policy makers as they make decisions in various countries on issues that involve human activities that affect the climate.

Climate variability and change are other areas requiring more research. Climatologists must understand short- and long-term climate change and how they are linked. Advances in computer modeling in recent years have enabled scientists to make great strides toward understanding the relationships of climate systems both geographically and over time. Researchers are focusing on predicting future changes on global scales as well as regional and local scales. As satellite imagery becomes more detailed and computer technology becomes more advanced, scientists are able to better understand, model, and project

FACTS ABOUT CLIMATE CHANGE

The Earth's climate has changed significantly over the last century. Scientists have found irrefutable evidence of global warming. As scientists study this evidence and use the information to build computer models, their goal is to be able to better understand how humans are not only affecting the Earth's climate today, but also how they will in the future. The many things they do know about climate change include:

- Current CO_2 levels are higher than anytime in the last 420,000 years.
- The Earth's global temperature has increased by 1°F (0.6°C) over the 1900s. Some areas have increased more than 3.3°F (2°C).
- The warmest year on record was 2006. The hottest years on record have all taken place within the last decade.
- Some scientists have predicted that greenhouse gas concentrations could rise as much as 950 ppm by the year 2100 (an increase of 670 ppm).
- Glacier National Park in Montana has had 110 glaciers completely disappear over the last 150 years. If the warming continues, the park could lose all of its glaciers in the next few decades.
- In 2003, a heat wave caused 35,000 people to die in Europe.
- Scientists predict sea levels will rise up to 20 feet (6 m) by the year 2150.
- The 2005 Atlantic hurricane season was the most active and destructive ever recorded.

Source: TheClimateGroup.org

the Earth's physical processes. Due to the complexities of the climate system, however, much research still needs to be done.

We must increase our knowledge of the global carbon cycle. The carbon cycle is complex, with many variables and relationships, both direct and indirect. Issues such as how the carbon reservoirs and carbon exchanges within the Earth operate, how carbon cycles through various components of the system, how long it resides in each component, how it is best managed from the human side of the cycle, and a better understanding of its long-term interactions with other components of the Earth's systems, such as the global water cycle, all need further research.

In order to further understanding of climate variability and the effects of global warming, it is also necessary to better understand the global water cycle. By understanding the intricate components of the water cycle, scientists will be able to comprehend its complexities (it involves chemical, physical, and biological processes that play a role in global warming and climate change). A better understanding will enable better modeling techniques, which will lead to better management strategies.

Jennifer Morgan, director of the World Wildlife Fund Climate Change Programme, says that "If we act now, it we address emissions now, we can avoid the worst case scenarios in the future. If it were too late, we would just work on adapting to climate change. We do some of that: we look at how we can make habitats more resilient, how you can extend protected areas, so that species can move. But the main goal for the world must be to cut down the emission of gas that pollutes the atmosphere and destroys the climate."

It is not too late to make a difference; to leave a livable world for future generations. While we will have to adapt to a certain degree to the damage that has already been done, much damage can be prevented by changing energy sources and lifestyle habits now. Planning and changing now while looking ahead will provide the key to a brighter tomorrow.

CHRONOLOGY

ca.1400–1850   Little Ice Age covers the Earth with record cold, large glaciers, and snow. There is widespread disease, starvation, and death.

1800–70   The levels of CO_2 in the atmosphere are 290 ppm.

1824   Jean-Baptiste Joseph Fourier, a French mathematician and physicist, calculates that the Earth would be much colder without its protective atmosphere.

1827   Jean-Baptiste Joseph Fourier presents his theory about the Earth's warming. At this time many believe warming is a positive thing.

1859   John Tyndall, an Irish physicist, discovers that some gases exist in the atmosphere that block infrared radiation. He presents the concept that changes in the concentration of atmospheric gases could cause the climate to change.

1894   Beginning of the industrial pollution of the environment.

1913–14   Svante Arrhenius discovers the greenhouse effect and predicts that the Earth's atmosphere will continue to warm. He predicts that the atmosphere will not reach dangerous levels for thousands of years, so his theory is not received with any urgency.

1920–25   Texas and the Persian Gulf bring productive oil wells into operation, which begins the world's dependency on a relatively inexpensive form of energy.

1934   The worst dust storm of the dust bowl occurs in the United States on what historians would later call Black Sunday. Dust storms are a product of drought and soil erosion.

1945   The U.S. Office of Naval Research begins supporting many fields of science, including those that deal with climate change issues.

1949–50   Guy S. Callendar, a British steam engineer and inventor, propounds the theory that the greenhouse effect is linked to human actions and will cause problems. No one takes him too seriously, but scientists do begin to develop new ways to measure climate.

1950–70   Technological developments enable increased awareness about global warming and the enhanced greenhouse effect. Studies confirm a steadily rising CO_2 level. The public begins to notice and becomes concerned with air pollution issues.

1958   U.S. scientist Charles David Keeling of the Scripps Institution of Oceanography detects a yearly rise in atmospheric CO_2. He begins collecting continuous CO_2 readings at an observatory on Mauna Loa, Hawaii. The results became known as the famous Keeling Curve.

1963   Studies show that water vapor plays a significant part in making the climate sensitive to changes in CO_2 levels.

1968   Studies reveal the potential collapse of the Antarctic ice sheet, which would raise sea levels to dangerous heights, causing damage to places worldwide.

1972   Studies with ice cores reveal large climate shifts in the past.

1974   Significant drought and other unusual weather phenomenon over the past two years cause increased concern about climate change not only among scientists but with the public as a whole.

1976   Deforestation and other impacts on the ecosystem start to receive attention as major issues in the future of the world's climate.

1977   The scientific community begins focusing on global warming as a serious threat needing to be addressed within the next century.

1979   The World Climate Research Programme is launched to coordinate international research on global warming and climate change.

1982   Greenland ice cores show significant temperature oscillations over the past century.

1983   The greenhouse effect and related issues get pushed into the political arena through reports from the U.S. National Academy of Sciences and the Environmental Protection Agency.

1984–90 The media begins to make global warming and its enhanced greenhouse effect a common topic among Americans. Many critics emerge.

1987   An ice core from Antarctica analyzed by French and Russian scientists reveals an extremely close correlation between CO_2 and temperature going back more than 100,000 years.

1988   The United Nations set up a scientific authority to review the evidence on global warming. It is called the Intergovernmental Panel on Climate Change (IPCC) and consists of 2,500 scientists from countries around the world.

1989   The first IPCC report says that levels of human-made greenhouse gases are steadily increasing in the atmosphere and predicts that they will cause global warming.

1990   An appeal signed by 49 Nobel prizewinners and 700 members of the National Academy of Sciences states, "There is broad agreement within the scientific community that amplification of the Earth's natural greenhouse effect by the buildup of various gases introduced by human activity has the potential to produce dramatic changes in climate . . . Only by taking action now can we insure that future generations will not be put at risk."

1992   The United Nations Conference on Environment and Development (UNCED), known informally as the Earth Summit, begins on June 3 in Rio de Janeiro, Brazil. It results in the United Nations Framework Convention on Climate Change, Agenda 21, the Rio Declaration on Environment and Development Statement of Forest Principles, and the United Nations Convention on Biological Diversity.

1993   Greenland ice cores suggest that significant climate change can occur within one decade.

1995   The second IPCC report is issued and concludes there is a human-caused component to the greenhouse effect warming. The consensus is that serious warming is likely in the coming century. Reports on the breaking up of Antarctic ice sheets and other signs of warming in the polar regions are now beginning to catch the public's attention.

1997   The third conference of the parties to the Framework Convention on Climate Change is held in Kyoto, Japan. Adopted on December 11, a document called the Kyoto Protocol commits its signatories to reduce emissions of greenhouse gases.

2000   Climatologists label the 1990s the hottest decade on record.

2001   The IPPC's third report states that the evidence for anthropogenic global warming is incontrovertible, but that its effects on climate are still difficult to pin down. President Bush declares scientific uncertainty too great to justify Kyoto Protocol's targets.

The United States Global Change Research Program releases the findings of the National Assessment of the Potential Consequences of Climate Variability and Change. The assessment finds that temperatures in the United States will rise by 5 to 9°F (3–5°C) over the next century and predicts increases in both very wet (flooding) and very dry (drought) conditions. Many ecosystems are vulnerable to climate change. Water supply for human consumption and irrigation is at risk due to increased probability of drought, reduced snow pack, and increased risk of flooding. Sea-level rise and storm surges will most likely damage coastal infrastructure.

2002   Second hottest year on record.

Heavy rains cause disastrous flooding in Central Europe leading to more than 100 deaths and more than $30 billion in damage. Extreme drought in many parts of the world (Africa, India, Australia, and the United States) results in thousands of deaths and significant crop damage. President Bush calls for 10 more years of research on climate change to clear up remaining uncertainties and proposes only voluntary measures to mitigate climate change until 2012.

2003   U.S. senators John McCain and Joseph Lieberman introduce a bipartisan bill to reduce emissions of greenhouse gases nation-wide via a greenhouse gas emission cap and trade program.

Scientific observations raise concern that the collapse of ice sheets in Antarctica and Greenland can raise sea levels faster than previously thought.

A deadly summer heat wave in Europe convinces many in Europe of the urgency of controlling global warming but does not equally capture the attention of those living in the United States.

International Energy Agency (IEA) identifies China as the world's second largest carbon emitter because of their increased use of fossil fuels.

The level of CO_2 in the atmosphere reaches 382 ppm.

2004   Books and movies feature global warming.

2005   Kyoto Protocol takes effect on February 16. In addition, global warming is a topic at the G8 summit in Gleneagles, Scotland, where country leaders in attendance recognize climate change as a serious, long-term challenge.

Hurricane Katrina forces the U.S. public to face the issue of global warming.

2006   Former U.S. vice president Al Gore's *An Inconvenient Truth* draws attention to global warming in the United States.

Sir Nicholas Stern, former World Bank economist, reports that global warming will cost up to 20 percent of worldwide gross domestic product if nothing is done about it now.

2007 IPCC's fourth assessment report says glacial shrinkage, ice loss, and permafrost retreat are all signs that climate change is underway now. They predict a higher risk of drought, floods, and more powerful storms during the next 100 years. As a result, hunger, homelessness, and disease will increase. The atmosphere may warm 1.8 to 4.0°C and sea levels may rise 7 to 23 inches (18 to 59 cm) by the year 2100.

Al Gore and the IPCC share the Nobel Peace Prize for their efforts to bring the critical issues of global warming to the world's attention.

2008 The price of oil reached and surpassed $100 per barrel, leaving some countries paying more than $10 per gallon.

Energy Star appliance sales have nearly doubled. Energy Star is a U.S. government-backed program helping businesses and individuals protect the environment through superior energy efficiency.
U.S. wind energy capacity reaches 10,000 megawatts, which is enough to power 2.5 million homes.

2009 President Obama takes office and vows to address the issue of global warming and climate change by allowing individual states to move forward in controlling greenhouse gas emissions. As a result, American automakers can prepare for the future and build cars of tomorrow and reduce the country's dependence on foreign oil. Perhaps these measures will help restore national security and the health of the planet, and the U.S. government will no longer ignore the scientific facts.

The year 2009 will be a crucial year in the effort to address climate change. The meeting on December 7–18 in Copenhagen, Denmark, of the UN Climate Change Conference promises to shape an effective response to climate change. The snapping of an ice bridge in April 2009 linking the Wilkins Ice Shelf (the size of Jamaica) to Antarctic islands could cause the ice shelf to break away, the latest indication that there is no time to lose in addressing global warming.

GLOSSARY

adaptation An adjustment in natural or human systems to a new or changing environment. Adaptation to climate change refers to adjustments in natural or human systems in response to actual or expected climatic changes.

aerosols Tiny bits of liquid or solid matter suspended in air. They come from natural sources such as erupting volcanoes and from waste gases emitted from automobiles, factories, and power plants. By reflecting sunlight, aerosols cool the climate and offset some of the warming caused by greenhouse gases.

albedo The relative reflectivity of a surface. A surface with high albedo reflects most of the light that shines on it and absorbs very little energy; a surface with a low albedo absorbs most of the light energy that shines on it and reflects very little.

anthropogenic Made by people or resulting from human activities. This term is usually used in the context of emissions that are produced as a result of human activities.

atmosphere The thin layer of gases that surround the Earth and allow living organisms to breathe. It reaches 400 miles (644 km) above the surface, but 80 percent is concentrated in the troposphere—the lower seven miles (11 km) above the Earth's surface.

biodiversity Different plant and animal species.

biomass Plant material that can be used for fuel.

bleaching (coral) The loss of algae from corals that causes the corals to turn white. This is one of the results of global warming and signifies a dieoff of unhealthy coral.

carbon A naturally abundant nonmetallic element that occurs in many inorganic and in all organic compounds.

carbon cycle A biogeochemical cycle in which carbon is exchanged among the biosphere, pedosphere, geosphere, hydrosphere, and atmosphere of Earth.

carbon dioxide   A colorless, odorless gas that passes out of the lungs during respiration. It is the primary greenhouse gas and causes the greatest amount of global warming.

carbon sink   An area where large quantities of carbon are built up in the wood of trees, in calcium carbonate rocks, in animal species, in the ocean, or any other place where carbon is stored. These places act as reservoirs, keeping carbon out of the atmosphere.

chlorofluorocarbons (CFCs)   Gases that were once widely used as coolants in refrigerators and air conditioners, as foaming agents for insulation and food packaging, and as cleaning agents in certain industries. They are long-lasting compounds that absorb heat energy more effectively than CO_2. When they enter the upper atmosphere, they destroy ozone (which protects life on Earth from harmful ultraviolet radiation). An international treaty calls for all production of CFCs to stop by the year 2010.

climate   The usual pattern of weather that is averaged over a long period of time.

climate feedback   An interaction between processes in the climate system is called a climate feedback, when the result of an initial process triggers changes in a second process that in turn influences the initial one. A positive feedback intensifies the original process, and a negative feedback reduces it.

climate model   A quantitative way of representing the interactions of the atmosphere, oceans, land surface, and ice. Models can range from relatively simple to extremely complicated.

climatologist   A scientist who studies the climate.

concentration   The amount of a component in a given area or volume. In global warming, it is a measurement of how much of a particular gas is in the atmosphere compared to all of the gases in the atmosphere.

condense   The process that changes a gas into a liquid.

Coriolis force   The effect caused by the spherical shape of the Earth that results in winds veering to the right in the Northern Hemisphere and to the left in the Southern Hemisphere.

cyclone   A large-scale, atmospheric wind and pressure system characterized by low pressure at the center and circular wind motion, counterclockwise in the Northern Hemisphere, clockwise in the Southern Hemisphere.

deflection   The angle formed by the line of sight to the target and the line of sight to the point at which an object is aimed. Because of the Earth's rotation, an object set into motion through the atmosphere will appear to deflect, or curve, from its beginning point to ending point in the path of travel because the Earth has moved underneath it while it was airborne.

deforestation   The large-scale cutting of trees from a forested area, often leaving bare areas susceptible to erosion.

doldrums   A belt of calms and light winds north of the equator between the northern and southern trade winds in the Atlantic and Pacific Oceans.

downwelling   A downward current of surface water in the ocean, usually caused by differences in the density of seawater.

ecosystem   A community of interacting organisms and their physical environment.

El Niño   A cyclic weather event in which the waters of the eastern Pacific Ocean off the coast of South America become much warmer than normal and disturb weather patterns across the region. Every few years, the temperature of the western Pacific rises several degrees above that of waters to the east. The warmer water moves eastward, causing shifts in ocean currents, jet stream winds, and weather in both the Northern and Southern Hemispheres.

emissions   The release of a substance (usually a gas when referring to the subject of climate change) into the atmosphere.

emissivity   The ability of a surface to emit radiant energy compared to that of a black body at the same temperature and with the same area.

evaporation   The process by which a liquid, such as water, is changed to a gas.

feedback   A change caused by a process that, in turn, may influence that process. Some changes caused by global warming may hasten the process of warming (positive feedback); some may slow warming (negative feedback).

forcings   Mechanisms that disrupt the global energy balance between incoming energy from the Sun and outgoing heat from the Earth. By altering the global energy balance, such mechanisms force the climate to change. Today, anthropogenic greenhouse gases added to the atmosphere are forcing climate to change.

fossil fuel   An energy source made from coal, oil, or natural gas. The burning of fossil fuels is one of the chief causes of global warming.

glacier   A mass of ice formed by the buildup of snow over hundreds and thousands of years.

global dimming   A reduction in the amount of the Sun's electromagnetic energy reaching the Earth's surface due to its blockage by particulate matter, clouds, and other opaque materials in the atmosphere.

global warming   An increase in the temperature of the Earth's atmosphere, caused by the buildup of greenhouse gases. This is also referred to as the enhanced greenhouse effect caused by humans.

global warming potential (GWP)   The cumulative radiative forcing effects of a gas over a specified time resulting from the emission of a unit mass of gas relative to a reference gas (usually CO_2).

great ocean conveyor belt   A global current system in the ocean that transports heat from one area to another.

greenhouse effect   The natural trapping of heat energy by gases present in the atmosphere, such as CO_2, methane, and water vapor. The trapped heat is then emitted as heat back to the Earth.

greenhouse gas   A gas that traps heat in the atmosphere and keeps the Earth warm enough to allow life to exist.

Gulf Stream   A warm current that flows from the Gulf of Mexico across the Atlantic Ocean to northern Europe. It is largely responsible for Europe's milder climate.

halogens   Any of a group of five nonmetallic elements with similar properties. The halogens are fluorine, chlorine, bromine, iodine, and astatine. Because they are missing an electron from their outermost shell, they react readily with most metals to form salts.

hydrologic cycle   The natural sequence through which water passes into the atmosphere as water vapor, precipitates to earth in liquid or solid form, and ultimately returns to the atmosphere through evaporation.

Industrial Revolution The period during which industry developed rapidly as a result of advances in technology. This took place in Britain during the late 18th and early 19th centuries.

infrared The invisible heat radiation that is emitted by the Sun and by virtually every warm substance or object on Earth.

IPCC Intergovernmental Panel on Climate Change. This is an organization consisting of 2,500 scientists that assesses information in the scientific and technical literature related to the issue of climate change. The IPCC was established jointly by the United Nations Environment Programme and the World Meteorological Organization in 1988.

jet stream A strong ribbon of horizontal wind that is found about 6 to 10 miles (10–16 km) above the ground in the area between the troposphere, the lower layer of the atmosphere, and the stratosphere above it.

Keeling Curve A famous curve showing increasing CO_2 concentrations in the atmosphere, which was set up by Dr. Charles David Keeling of Scripps Institution of Oceanography at Mauna Loa in Hawaii, it illustrates the steady rise in CO_2 concentrations since 1958.

land breeze Opposite of a sea breeze, a breeze that blows from the land toward open water.

land use The management practice of a certain land cover type, forest, arable land, grassland, urban land, and wilderness.

land use change An alteration of the management practice on a certain land cover type. Land use changes may influence climate systems because they influence evapotranspiration and sources and sinks of greenhouse gases, i.e., removing a forest to build a city.

mass wasting The downslope movement of rock and regolith near the Earth's surface mainly due to the force of gravity.

Maunder minimum The period of reduced solar activity lasting through the 1600s and 1700s.

mesosphere The layer of the atmosphere that lies on top of the stratosphere and runs upward to about 50 miles (80 km).

methane　A colorless, odorless, flammable gas that is the major ingredient of natural gas. Methane is produced wherever decay occurs and little or no oxygen is present.

monsoon　Heavy rains that occur at the same time each year.

mountain breeze　It is formed at night by the radiational cooling along mountainsides. As the slopes become colder than the surrounding atmosphere, the lower levels of air cool and drain to the lowest point of the terrain. It may reach several hundred feet in depth and, in extreme cases, attain speeds of 57.5 miles per hour (50 knots) or greater. It blows in the opposite direction of a valley breeze.

nitrogen　As a gas, nitrogen takes up 80 percent of the volume of the Earth's atmosphere. It is also an element in substances such as fertilizer.

nitrous oxide　A heat-absorbing gas in the Earth's atmosphere. Nitrous oxide is emitted from nitrogen-based fertilizers.

nuclear power　The electricity produced by a process that begins with the splitting apart of uranium atoms, yielding great amounts of heat energy.

ocean ridges　An area under the ocean where the Earth's crust is pulling apart, allowing new lava to emerge on the ocean floor, harden, and become new crust in the plate tectonic process.

ozone　A molecule that consists of three oxygen atoms. Ozone is present in small amounts in the Earth's atmosphere at 14 to 19 miles (23–31 km) above the Earth's surface. A layer of ozone makes life possible by shielding the Earth's surface from most harmful ultraviolet rays. In the lower atmosphere, ozone emitted from auto exhausts and factories is an air pollutant.

paleomagnetism　The study of the record of the Earth's magnetic field preserved in various magnetic minerals through time.

parts per million (ppm)　The number of parts of a chemical found in one million parts of a particular gas, liquid, or solid.

permafrost　Permanently frozen ground in the Arctic. As global warming increases, this ground is melting.

photosynthesis　The process by which plants make food using light energy, carbon dioxide, and water.

protocol   The terms of a treaty that have been agreed to and signed by all parties.

proxies   Methods of determining values such as temperatures and rainfall by using substitutes, which give indirect measurements. Tree rings serve as proxies for determining rainfall abundance.

radiation   The particles or waves of energy.

renewable   Something that can be replaced or regrown, such as trees, or a source of energy that never runs out, such as solar energy, wind energy, or geothermal energy.

resources   The raw materials from the Earth that are used by humans to make useful things.

rotation   The movement or path of the Earth, turning on its axis.

satellite   Any small object that orbits a larger one. Artificial satellites carry instruments for scientific study and communication. Imagery taken from satellites is used to monitor aspects of global warming such as glacier retreat, ice cap melting, desertification, erosion, hurricane damage, and flooding. Sea surface temperatures and measurements are also obtained from man-made satellites in orbit around the Earth.

sea breeze   A wind from the sea that develops over land near coasts. It is formed by increasing temperature differences between the land and water that create a pressure minimum over the land due to its relative warmth and forces higher pressure, cooler air from the sea to move inland.

simulation   A computer model of a process that is based on actual facts. The model attempts to mimic, or replicate, actual physical processes.

stratosphere   The layer of the atmosphere just above the troposphere. It extends 7.5 miles (12 km) to an average of 31 miles (50 km).

temperate   An area that has a mild climate and different seasons.

thermal   Something that relates to heat.

trace gases   Gases found in minute amounts in the atmosphere.

trade winds   Winds that blow steadily from east to west and toward the equator. The trade winds are caused by hot air rising at the equator, with cool air moving in to take its place from the north and from the south. The winds are deflected westward because of the Earth's west-to-east rotation.

tropical   A region that is hot and often wet (humid). These areas are located around the Earth's equator.

tropical depression   A cyclone in a tropical region that is more intense than a disturbance but less than a storm, with wind speeds of 38 mph (61 km/h) or less.

tropical disturbance   A very weak tropical cyclone.

tropical storm   A cyclonic storm having winds ranging from approximately 30 to 75 miles (48–121 kilometers) per hour.

troposphere   The bottom layer of the atmosphere, rising from sea level up to an average of about 7.5 miles (12 km).

tundra   A vast treeless plain in the Arctic with a marshy surface covering a permafrost layer.

ultraviolet radiation   A portion of the Sun's electromagnetic spectrum, consisting of very short wavelengths and high energy. The atmosphere's ozone layer protects life on Earth from the damaging effects from UV radiation.

upwelling   The process by which warm, less-dense surface water is drawn away from along a shore by offshore currents and replaced by cold, denser water brought up from the subsurface.

valley breeze   An anabatic wind, it is formed during the day by the heating of the valley floor. As the ground becomes warmer than the surrounding atmosphere, the lower levels of air heat and rise, flowing up the mountainsides. It blows in the opposite direction of a mountain breeze.

visible light   The wavelengths of the Sun's electromagnetic spectrum that humans can see; it falls in the wavelength range of 400 to 700 nm.

weather   The conditions of the atmosphere at a particular time and place. Weather includes such measurements as temperature, precipitation, air pressure, and wind speed and direction.

weathering   The progression of breaking down rocks and natural materials on the Earth's surface through physical and chemical processes.

westerlies   A semipermanent belt of westerly winds that prevails at latitudes lying between the tropical and polar regions of the Earth.

FURTHER RESOURCES

BOOKS

Christianson, Gale. *Greenhouse: The 200-Year Story of Global Warming.* New York: Walker, 1999. Examines the enhanced greenhouse effect worldwide after the industrial revolution and outlines the consequences to the environment.

Cox, John D. *Climate Crash—Abrupt Climate Change and What It Means for Our Future.* Washington, D.C: Joseph Henry Press, 2005. Outlines the science of paleoclimatology and how events of the past hold clues to the present and future.

Flannery, Tim. *The Weather Makers.* New York: Atlantic Monthly Press, 2005. Explores how humans are changing the climate and how it will affect all life on Earth.

Friedman, Katherine. *What If the Polar Ice Caps Melted?* Danbury, Conn.: Children's Press, 2002. Focuses on environmental problems related to the Earth's atmosphere, including global warming, changing weather patterns, and their effects on ecosystems.

Gelbspan, Ross. *The Heat Is On: The High Stakes Battle over Earth's Threatened Climate.* Reading, Mass.: Addison Wesley, 1997. Looks at the controversy environmentalists often face when they deal with fossil fuel companies.

——. *Boiling Point.* New York: Basic Books, 2004. Presents the role that politicians, oil companies, the media, and activists have had in shaping people's beliefs about the issue of global warming.

Gore, Albert. *An Inconvenient Truth.* Emmaus, Penn.: Rodale, 2006. Presents an excellent overview of the global warming problem, how it has come about, what it means for the future, and why humans need to act now to slow it down.

Hamblin, W. Kenneth, and Eric H. Christiansen. *Earth's Dynamic Systems.* Tenth ed. Upper Saddle River, N.J.: Prentice Hall, Inc. Discusses the physical aspects of Earth science, explaining the physical mechanisms behind plate tectonics, glaciers, shorelines, groundwater, river systems, and other systems that are affected by global warming.

Harrison, Patrick, Gail McLeod, and Patrick G. Harrison. *Who Says Kids Can't Fight Global Warming.* Chattanooga, Tenn.: Pat's Top Products, 2007. Offers real solutions to help solve the world's biggest air pollution problems.

Houghton, John. *Global Warming: The Complete Briefing.* New York: Cambridge University Press, 2004. Outlines the scientific basis of global warming and describes the impacts that climate change will have on society. It also looks at solutions to the problem.

Langholz, Jeffrey. *You Can Prevent Global Warming (and Save Money!): 51 Easy Ways.* Riverside, N.J.: Andrews McMeel Publishing, 2003. Aims at converting public concern over global warming into positive action to stop it by providing simple, everyday practices that can easily be done to minimize it, as well as save money.

Leggett, Jeremy. *The Carbon War.* New York: Routledge, 2001. Presents the political view of global warming and the conflict with the oil industry.

Linden, Eugene. *The Winds of Change.* New York: Simon & Schuster, 2006. Looks at past and present climate change and what the future holds.

Lynas, Mark. *High Tide.* New York: Picador, 2004. Explores the global effects of climate change on the Earth's diverse ecosystems.

McKibben, Bill. *Fight Global Warming Now: The Handbook for Taking Action in Your Community.* New York: Holt, 2007. Provides the facts of what must change to save the climate. It also shows how everyone can act proactively in their community to make a difference.

Michaels, Patrick J. *Meltdown.* Washington, D.C.: Cato Institute, 2004. Discusses the issues of climate change as seen not only by scientists, but by politicians and the media as well.

———. *Shattered Consensus.* Lanham, Md.: Rowman & Littlefield, 2005. A series of essays from climate change experts on key issues surrounding global warming and interpretations of how severe the problem is.

National Research Council. *Abrupt Climate Change—Inevitable Surprises.* Washington D.C.: National Academy Press, 2002. Explores the causes and reality of abrupt climate change and its global effects.

Pearce, Fred. *When the Rivers Run Dry.* Boston: Beacon Press, 2006. Discusses the problem of drought as one of the major crises that will occur in the 21st century as a result of global warming.

Pringle, Laurence. *Global Warming: The Threat of Earth's Changing Climate.* New York: SeaStar Publishing Company, 2001. Provides information on the carbon cycle, rising sea levels, El Niño, aerosols, smog, flooding, and other issues related to global warming.

Stern, Nicholas. *The Economics of Climate Change.* New York: Cambridge University Press, 2007. This book discusses the economic ramifications of global warming.

Thornhill, Jan. *This Is My Planet—The Kids Guide to Global Warming.* Toronto, Ontario: Maple Tree Press, 2007. Offers students the tools they need to learn about ecology by taking a comprehensive look at climate change in polar, ocean, and land-based ecosystems.

Weart, Spencer R. *The Discovery of Global Warming (New Histories of Science, Technology, and Medicine).* Cambridge, Mass.:Harvard University Press, 2004. Traces the history of the global warming concept through a long process of incremental research rather than a dramatic revelation.

ARTICLES

Adam, David. "Climate Scientists Issue Dire Warning." *Guardian,* February 28, 2006. Presents new evidence concerning the ramifications of global warming.

Alley, Richard B. "Abrupt Climate Change." *Scientific American* 291, no. 5 (November 2004): 62–69. Discusses the great ocean conveyor belt and how global warming could shut it down and trigger another ice age.

Appenzeller, Tim. "The Big Thaw." *National Geographic* (June 2007): 58–71. An overview of global melting and what that means for the future.

Applebome, Peter. "A Community Tries to Shrink Its Footprint." *New York Times* (January 20, 2008). Available online. URL:

http://www.nytimes.com/2008/01/20/nyregion/20towns.html?_
r=2&partner=rssnyt<0x 0026>emc=rss&oref=slogin&oref=slogin.
Accessed September 21, 2008. Profiles Westport, Connecticut, and the
steps that it is taking to become more environmentally responsible.

Barringer, Felicity. "U.S. Given Poor Marks on the Environment." *New
York Times* (January 23, 2008). Available online. URL: http://www.
nytimes.com/2008/01/23/washington/23enviro.html. Accessed September 22, 2008. The United States lags behind on environmental
responsibility and positive action.

Bindschadler, Robert A., and Charles R. Bentley. "On Thin Ice?" *Scientific American* (December 2002): 98–105. Discusses the melting rates
of the Earth's most massive ice sheets and what controls the rates of
disintegration.

Borenstein, Seth. "Hotter Weather Linked to More Extinctions." *Discovery News* (October 24, 2007). Available online. URL: http://dsc.discov-
ery.com/news/2007/10/24/extinctions-warming-print.html. Accessed
September 17, 2008. Examines the Earth's past and how scientists
study extinctions, as well as what effects today's global warming may
have on present life on Earth.

Bosire, Bogonko. "More Than 1 Billion Trees Planted in 2007." *Discovery
News* (November 28, 2007). Available online. URL: http://dsc.discov-
ery.com/news/2007/11/28/tree-planting-forests-print.html. Accessed
September 17, 2008. Describes how more than 1 billion trees were
planted around the world in 2007, with Ethiopia and Mexico leading
in the drive to combat climate change.

Bourne, Joel K., Jr. "Green Dreams." *National Geographic* (October
2007): 38–59. Explores the use of biofuels.

Britt, Robert Roy. "Surprising Side Effects of Global Warming." Live-
Science (December 22, 2004). Available online. URL: http://www.live-
science.com/environment. Accessed September 17, 2008. Addresses
the damage to the landscape from melting permafrost, landslides, and
mudslides as a result of global warming.

————. "Global Warming Likely Cause of Worst Mass Extinction Ever."
LiveScience (January 20, 2005). Available online. URL: http://www.
livescience.com/environment/050120_great-dying.html. Accessed Sep-

tember 22, 2008. Presents scientific evidence that past extinctions were related to changing climate.

———. "Caution: Global Warming May Be Hazardous to Your Health." LiveScience (February 2005). Available online. URL: http://www.live-science.com/environment/050221_warming_heath.html. Accessed September 22, 2008. This article addresses the timely issues of pollution and the effects on human health.

———. "Energy Imbalance Behind Global Warming." LiveScience (April 28, 2005). Available online. URL: http://www.livescience.com/environment_050428/solar_energy.html. Accessed September 22, 2008. Looks at existing heat brought on by air pollution and slowly warming the Earth where it will become out of control.

———. "125 Large Northern Lakes Disappear." LiveScience (June 3, 2005). Available online. URL: http://www.livescience.com/environment/050603_lakes_gone.html. Accessed September 22, 2008. Focuses on lakes in the Arctic that have vanished as temperatures have risen over the past two decades and what effect this is having.

———. "Global Warming Sparks Increased Plant Production in Arctic Lakes." LiveScience (October 24, 2005). Available online. URL: http://www.livescience.com/environment/051024_arctic_lakes.html. Accessed September 17, 2008. Looks at longer growing seasons in the arctic ecosystems as a result of global warming.

———. "Insurance Company Warns of Global Warming's Costs." LiveScience (November 1, 2005). Available online. URL: http://www.live-science.com/environment/051101_insurance_warning.html. Accessed September 22, 2008. Explores the economic issues behind the destruction caused by global warming.

———. "The Irony of Global Warming: More Rain, Less Water," LiveScience (November 16, 2005). Available online. URL: http://www.livescience.com/environment/051116_water_shortage.html. Accessed September 22, 2008. Outlines that an enhanced water cycle may not benefit the necessary water supplies.

———. "Conflicting Claims on Global Warming and Why It's All Moot." LiveScience (February 1, 2006). Available online. URL: http://www.livescience.com/environment/060201_temperature_differences.html.

Accessed September 22, 2008. Presents the pros and cons of global warming.

Carey, Bjorn. "Soot Could Hasten Melting of Arctic Ice." LiveScience (March 28, 2005). Available online. URL: http://www.livescience.com/environment/050328_arctic-soot.html. Accessed September 22, 2008. Shows how when ice becomes polluted with soot, it changes the reflective properties and accelerates melting in the Arctic.

———. "Arctic Summer Could Be Ice-Free by 2105." LiveScience (August 23, 2005). Available online. URL: http://www.livescience.com/environment/050823_ice_free.html. Accessed September 22, 2008. Discusses the accelerated rates of ice melting.

Carlton, Jim. "Is Global Warming Killing the Polar Bears?" *Wall Street Journal* (December 14, 2005). Available online. URL: http://online.wsj.com/public/article_print/SB113452435089621905-vnekw47PQGtDy-f3iv5XE N71_o5I_20061214.html. Accessed September 22, 2008. Discusses the plight currently facing the polar bears and their survival.

Castle, Stephen, and James Kanter. "Stricter System to Trim Carbon Emissions is Considered in Europe." *New York Times* (January 22, 2008). Available online. URL: http://www.nytimes.com/2008/01/22/business/worldbusiness/22emissions.html. Accessed September 22, 2008. Explores the concept of reducing carbon emissions.

CBS News. "'Monumental' Climate Change?" *CBS News* (June 19, 2007). Available online. URL: http://www.cbsnews.com/stories/2007/06/19/eveningnews/main2952286.shtml. Accessed September 22, 2008. This article outlines the destruction on the world's fine art as a result of pollution and global warming.

Coleman, Joseph. "Seaweed: The New Carbon Sink?" *Discovery News* (December 10, 2007). Available online. URL: http://dsc.discovery.com/news. Accessed September 22, 2008. This article looks at the world's carbon sinks and their potential to offset the effects of global warming.

Culotta, Elizabeth. "Will Plants Profit from High CO_2?" *Science* (May 5, 1995): 654–656. Explores the effects of various CO_2 levels on vegetation as a result of global warming and whether they will experience enhanced growth.

D'Agnese, Joseph. "Why Has Our Weather Gone Wild?" *Discover* (June 2000): 72–78. Focuses on the recent global changes in weather, such as shifting seasons, severe storms, droughts, and heat waves, and their connection to global warming.

Davidson, Sarah. "How Global Warming Can Chill the Planet." LiveScience (December 17, 2004). Available online. URL: http://www.livescience.com/environment/041217_sealevel_rise.html. Accessed September 22, 2008. Explores how ocean's currents can lead to a cooler climate if they are disrupted through the addition of freshwater from melting ice caps and glaciers.

Dean, Cornelia. "The Preservation Predicament." *New York Times* (January 29, 2008). Available online. URL: http://www.nytimes.com/2008/01/29/science/earth/29habi.html?_r=1&ref=science<0x0026>oref=slogin. Accessed September 22, 2008. Looks at humans' current management of the landscape and wildlife on it and how habitats are going to survive.

Deutsch, Claudia H. "A Threat So Big, Academics Try Collaboration." *New York Times* (December 25, 2007). Available online. URL: http://www.nytimes.com/2007/12/25/business/25sustain.html. Accessed September 22, 2008. Explores the methodology of scientists working together to deal with an immense global issue.

Eilperin, Juliet. "Severe Hurricanes Increasing, Study Finds." *Washington Post* (September 16, 2005). Available online. URL: http://www.washingtonpost.com/wp-dyn/content/article/2005/09/15/AR2005091502234.html. Accessed September 22, 2008. Links rising sea temperatures to an increase in more destructive hurricanes.

Epstein, Paul R. "Is Global Warming Harmful to Health?" *Scientific American* (August 2000): 50–57. Explores the issues of widespread epidemics as a fallout of increased global warming.

Fountain, Henry. "Katrina's Damage to Trees May Alter Carbon Balance." *New York Times* (November 20, 2007). Available online. URL: http://www.nytimes.com/2007/11/20/science/20obtree.html. Accessed September 22, 2008. Looks at the ecological implications of the destruction of hundreds of trees on the Gulf Coast as a result of Hurricane Katrina.

————. "More Acidic Ocean Hurts Reef Algae as Well as Corals." *New York Times* (January 8, 2008). Available online. URL: http://www.nytimes.com/2008/01/08/science/earth/08obalga.html. Accessed September 22, 2008. As more carbon dioxide is added to the atmosphere, the oceans are becoming acidic and scientists believe they are damaging not only coral, but also reef algae.

Friedman, Thomas L. "It's Too Late for Later." *New York Times* (December 16, 2007). Available online. URL: http://www.nytimes.com/2007/12/16/opinion/16friedman.html. Accessed September 22, 2008. United Nations specialists and other leaders are warning that action must be taken immediately on global warming.

Gelling, Peter. "Focus of Climate Talks Shifts to Helping Poor Countries Cope." *New York Times* (December 13, 2007). Available online. URL: http://www.nytimes.com/2007/12/13/world/13climate.html. Accessed September 22, 2008. Outlines several courses of realistic adaptation for countries faced with the negative effects of global warming.

Gibbs, Walter, and Sarah Lyall. "Gore Shares Peace Prize for Climate Change Work." *New York Times* (October 13, 2007). Available online. URL: http://www.nytimes.com/2007/12/13/world/13climate.html. Accessed September 22, 2008. Former vice president Al Gore was awarded the 2007 Nobel Peace Prize on Friday, sharing it with the Intergovernmental Panel on Climate Change.

Goudarzi, Sara. "Allergies Getting Worse Due to Global Warming." LiveScience (November 22, 2005). Available online. URL: http://www.livescience.com/environment/051122_allergy_rise.html. Accessed September 22, 2008. Looks at how a warmer atmosphere can increase the occurrence of allergic systems in humans.

Hansen, James. "Defusing the Global Warming Time Bomb." *Scientific American* (March 2004): 68–77. Discusses the widespread scale of the problem today and why immediate action is urgently needed.

Heilprin, John. "Study: Earth Is Hottest Now in 2,000 Years; Humans Responsible for Much of the Warming." *USA Today* (June 23, 2006). Available online. URL: http://www.nytimes.com/2007/12/13/world/13climate.html?partner=rssnyt&emc=rss<0x0026>pagewanted=all. Accessed September 22, 2008. Discusses paleoclimatology and proxy evidence, as well as correlates past evidence to today's intense storms, heat waves, and other climate conditions.

Hennessey, Kathleen. "Dairy Air: Scientists Measure Cow Gas." Live-Science (July 27, 2005). Available online. URL: http://www.livescience. com/animals/ap_050727_cow_gas.html. Accessed September 22, 2008. A study that looks at agricultural practices and the connections to global warming.

Hertsgaard, Mark. "On the Front Lines of Climate Change." *Time* (April 9, 2007): 102–109. Explores the concepts of human adaptation to climate change.

Hoffman, Allison, and Gillian Flaccus. "Wildfires Engulf Southern California." Discovery News (October 22, 2007). Available online. URL: http://dsc.discovery.com/news/2007/10/22/wildfires-california-print. html. Accessed September 17, 2008. This article links wildfires with global warming and climate change.

Hoffman, Paul F., and Daniel P. Schrag. "Snowball Earth." *Scientific American* (January 2000): 68–75. Discusses the Earth when it was covered with ice millions of years ago.

Joling, Dan. "Study: Polar Bears May Turn to Cannibalism." *USA Today* (June 13, 2006). Available online. URL: http://www.usatoday.com/ weather/research/2006-06-13-polar-bear-cannibalism_x.htm. Accessed September 22, 2008. Discusses how the melting polar ice is keeping polar bears from being able to hunt for food in their traditional hunting grounds.

Kahn, Joseph, and Mark Landler. "China Grabs West's Smoke-Spewing Factories." *New York Times* (December 21, 2007). Available online. URL: http://www.nytimes.com/2007/12/21/world/asia/21transfer.html. Accessed September 22, 2008. This article looks at the energy revolution taking place in China.

Kahn, Joseph. "Japan Urges China to Reduce Pollution." *New York Times* (December 19, 2007). Available online. URL: http://topics.nytimes. com/top/reference/timestopics/people/f/yasuo_fukuda/index.html.

Accessed September 23, 2008. This article focuses on the political implications of global warming between countries.

Kanter, James. "Europe May Ban Imports of Some Biofuel Crops." *New York Times* (January 15, 2008). Available online. URL: www. nytimes.com/2008/01/15/business/worldbusiness/15biofuel.html?_ r=1&oref=slogin. Accessed September 23, 2008. Discusses a law that would prohibit the importation of fuels derived from crops grown on certain kinds of land.

Karl, Thomas R., Neville Nicholls, and Jonathan Gregory. "The Coming Climate." *Scientific American* (May 1997): 78–83. Discusses computer models and how they interpret weather patterns of a warmer world.

Karl, Thomas, and Kevin Trenberth. "The Human Impact on Climate." *Scientific American* (December 1999): 100–105. Focuses on the disruptions people cause in the natural environment and why scientists must begin to monitor and quantify the disruptions now in order to save the future.

Kaufman, Marc. "Research Shines Some Light on Mysteries of Antarctica." *Washington Post* (February 18, 2007). Available online. URL: http://www.washingtonpost.com/wp-dyn/content/article/2007/02/17/ AR2007021701335.html. Accessed September 23, 2008. Looks at snowfall, ice melt, and wind systems and how the three components work together in Antarctica.

Kay, Jane. "A Warming World: Climate Change Report." *San Francisco Chronicle* (February 3, 2007). Available online. URL: http://www. sfgate.com/cgi-bin/article.cgi?file=/c/a/2007/02/03/MNGR9NUBLN1. DTL. Accessed September 9, 2008. Focuses on the new report from the IPCC and their warnings of the effects of global warming to come.

Kerr, Jennifer C. "2007 Among Warmest Years Ever in U.S." Discovery News (December 14, 2007). Available online. URL: http://dsc. discovery.com/news/2007/12/14/warmest-year-weather.html. Accessed September 23, 2008. Discusses the events of 2007 and the role global warming may have played.

Kluger, Jeffrey. "A Climate of Despair: Special Report, Global Warming." *Time* (April 9, 2001): 30–36. Discusses the political ramifications of global warming and the need for global cooperation.

———. "Is Global Warming Fueling Katrina?" *Time* (August 29, 2005). Available online. URL: http://www.time.com/time/printout/0,8816,1099102,00.html. Accessed September 23, 2008. Looks at global warming as a cause of strengthening hurricanes.

———. "Global Warming Heats Up." *Time* (March 26, 2006). Available online. URL: http://www.time.com/time/printout/0,8816,1176980,00.html. Accessed September 23. 2008. Discusses several environmental aspects of the environment that point to the effects of global warming.

———. "What Now?" *Time* (April 9, 2007): 52–60. Discusses the problem from an international aspect and how everyone can be involved in the solution of the problem.

Krauss, Clifford. "As Ethanol Takes Its First Steps, Congress Proposes a Giant Leap." *New York Times* (December 18, 2007). Available online. URL: http://www.nytimes.com/2007/12/18/washington/18ethanol.html?n=Top/News/Science/Topics/ Global%20Warming. Accessed September 23, 2008. Discusses how the U.S. Congress is on the verge of writing into law one of the most ambitious dictates ever issued to American business: To create a new industry capable of converting agricultural wastes and other plant material into automotive fuel.

Lemonick, Michael D. "Life in the Greenhouse." *Time* (April 9, 2001): 24–29. Outlines the current signs of global warming and the effects it will have on the environment.

———. "Why You Can't Ignore the Changing Climate." Parade.com (June 25, 2006). Available online. URL: http://www.parade.com/articles/editions/2006/edition_06-25-2006. Accessed September 23, 2008. Points out that evidence of climate change is everywhere and not only needs to be taken seriously, but everyone needs to do their part to help fix it.

LiveScience. "Global Warming Could Overwhelm Storm Drains." (October 11, 2005). Available online. URL: http://www.livescience.com/environment/051011_culverts.html. Accessed September 23, 2007.

Social and management aspects of flooding during extreme storm events.

———. "Darker Days in China as Sun Gets Dimmer." (January 20, 2006). Available online. URL: http://www.livescience.com/ environment/060119_dark_china.html. Accessed September 23, 2007. This article discusses the hazards of increased fossil fuel use and pollution.

Marshall, Carolyn. "San Francisco Fleet Is All Biodiesel." *New York Times* (December 2, 2007). Available online. URL: http://www.nytimes. com/2007/12/02/us/02diesel.html. Accessed September 23, 2008. Discusses how San Francisco has converted all its diesel vehicles to biodiesel in order to reduce greenhouse gas emissions.

———. "Carbon's New Math." *National Geographic* (October, 2007): 32–37. Explores ways to use technology to reduce the effects of global warming.

Mouawad, Jad. "OPEC Gathering Finds High Oil Prices More Worrisome Than Welcome." *New York Times* (November 17, 2007). Available online. URL: http://www.nytimes.com/2007/11/17/ business/17opec.html. Accessed September 23, 2008. This article discusses the economic realities of climbing oil prices and issues concerning a global economic recession.

———. "Oil Demand, the Climate and the Energy Ladder." *New York Times* (January 19, 2008). Available online. URL: http://www.nytimes. com/2008/01/19/business/19interview.html. Accessed September 23, 2008. An article discussing the future energy demands and the need for setting limits on carbon emissions.

Neergaard, Lauran. "Greenhouse Gas at 650,000-year High." LiveScience (November 25, 2005). Available online. URL: http://www.livescience. com/environment/ap_051125_greenhouse_gas.html. Accessed September 23, 2008. Looks at the European Project for Ice Coring in Antarctica and relates evidence found by scientists in the ice cores to the conditions humans face today.

New York Times. "Connecting the Global Warming Dots." (January 14, 2007). Available online. URL: http://www.nytimes.com/2007/01/14/

weekinreview/14basics.html. Accessed September 23, 2008. Presents several reasons why global warming is occurring.

Nicklen, Paul. "Life at the Edge." *National Geographic* (June 2007): 32–55. Explores the shrinking sea ice at the North Pole and its effects on the ecosystem.

O'Hanlon, Larry. "U.S. Groundwater Drying Up." *Discovery News* (October 25, 2007). Available online. URL: http://dsc.discovery.com/news/2007/10/25/ground-water-drought-print.html. Accessed September 23, 2008. Discusses the impacts global warming is having on the United States and the supplies of water that are necessary to support communities.

———. "Desalinated Water: Great to Drink, Bad for Crops." *Discovery News* (November 8, 2008). Available online. URL: http://dsc.discovery.com/news/2007/11/08/desalination-agriculture-print.html. Accessed September 23, 2008. Discusses the issues surrounding desalination of ocean water in view of combating water shortages.

Revkin, Andrew C. "Climate Experts Warn of More Coastal Building." *New York Times* (July 25, 2006). Available online. URL: http://www.nytimes.com/2006/07/25/science/earth/25coast.html. Accessed September 23, 2008. Explores the economic and social issues of building in coastal areas prone to hurricane damage.

———. "Global Warming Trend Continues in 2006, Climate Agencies Say." *New York Times* (December 15, 2006). Available online. URL: http://www.nytimes.com/2006/12/15/science/15climate.html. Accessed September 23, 2008. Reports on new record-setting heat waves worldwide.

———. "A New Middle Stance Emerges in Debate over Climate." *New York Times* (January 1, 2007). Available online. URL: http://www.nytimes.com/2007/01/01/science/01climate.html. Accessed September 23, 2008. Introduces a new middle-of-the-road group representing a more moderate view on global warming.

———. "U.S. Predicting Steady Increase for Emissions." *New York Times* (March 3, 2007). Available online. URL: http://www.nytimes.com/2007/03/03/science/03climate.html. Accessed September 23,

2008. Discusses information recently released from the White House predicting that emissions of carbon dioxide and other greenhouse gases are expected to increase over the next decade.

————. "Many Arctic Plants Have Adjusted to Big Climate Changes, Study Finds." *New York Times* (June 15, 2007). Available online. URL: http://www.nytimes.com/2007/06/15/science/15arctic.html. Accessed September 23, 2008. Discusses the concept that some plants may be able to shift long distances to follow the climate conditions for which they are best adapted as those conditions move under the influence of global warming.

————. "Arctic Melt Unnerves the Experts." *New York Times* (October 2, 2007). Available online. URL: http://www.nytimes.com/2007/10/02/science/earth/02arct.html. Accessed September 23, 2008. Presents information confirming that Arctic ice is melting faster than scientists originally expected and the global consequences of it.

————. "As China Goes, So Goes Global Warming." *New York Times* (December 16, 2007). Available online. URL: http://www.nytimes.com/2007/12/16/weekinreview/16revkin.html. Accessed September 23, 2008. Addresses the issue of clean energy for China.

————. "Connecting the Global Warming Dots." *New York Times* (January 14, 2007). Available online. URL: http://www.nytimes.com/2007/01/14/weekinreview/14basics.html. Accessed September 23, 2008. Presents the basics of what global warming is and why it is happening.

————. "Issuing a Bold Challenge to the U.S. over Climate." *New York Times* (January 22, 2008). Available online. URL: http://www.nytimes.com/2008/01/22/science/earth/22conv.html. Accessed September 23, 2008. Explores the international political arena and the pressure being put on the United States to become more involved with addressing global warming issues.

Rosenthal, Elisabeth, and Andrew C. Revkin. "Panel Issues Bleak Report on Climate Change." *New York Times* (February 2, 2007). Available online. URL: http://www.nytimes.com/2007/02/02/science/

earth/02cnd-climate.html. Accessed September 23, 2008. Review of the IPCC's newly issued report on global warming.

Rosenthal, Elisabeth. "Science Panel Calls Global Warming 'Unequivocal,'" *New York Times* (February 3, 2007). Available online. URL: http://www.nytimes.com/2007/02/03/science/earth/03climate.html. Accessed September 23, 2008. Profiles the IPCC's latest report on global warming.

———. "U.N. Report Describes Risks of Inaction on Climate Change." *New York Times* (November 17, 2007). Available online. URL: http://www.nytimes.com/2007/11/17/science/earth/17climate.html. Accessed September 23, 2008. Focuses on the attention the United Nations is trying to gain in order to get individual countries to contribute toward the solution of global warming.

———. "U.N. Chief Seeks More Climate Change Leadership." *New York Times* (November 18, 2007). Available online. URL: http://www.nytimes.com/2007/11/18/science/earth/18climatenew.html. Accessed September 23, 2008. Reviews the necessity of effective political leadership in order to back up the effort to stop global warming if it is to succeed.

Schirber, Michael. "Nature's Wrath: Global Deaths and Costs Swell." LiveScience (November 1, 2004). Available online. URL: http://www.livescience.com/environment/041101_disaster_report.html. Accessed August 13, 2008. Examines why weather-related disasters are more prominent now than they were.

———. "Global Disaster Hotspots: Who Gets Pummeled." LiveScience (December 7, 2004). Available online. URL: http://www.livescience.com/environment/050107_disaster_hotspots.htm. Accessed September 23, 2008. Discusses the most likely areas in the world to be negatively affected by natural disasters.

———. "Longer Airline Flights Proposed to Combat Global Warming." LiveScience (January 26, 2005). Available online. URL: http://www.livescience.com/environment/050126_contrail_climate.html. Accessed September 23, 2008. Investigates plans to limit the effects of contrails in the atmosphere from airline traffic.

ScienceDaily. "From Icehouse to Hothouse: Melting Ice and Rising Carbon Dioxide Caused Climate Shift." (February 27, 2007). Available online. URL: http://www.sciencedaily.com/releases/2007/02/0702211358.htm. Accessed September 23, 2008. Discusses modeling of paleoclimates in order to understand changes happening today.

Schmid, Randolph E. "Greenhouse Gas Hits Record High." LiveScience (March 15, 2006). Available online. URL: http://www.livescience.com/environment/ap_060315_carbon_dioxide.htm. Accessed September 23, 2008. Discusses how the measurements of atmospheric carbon dioxide have changed, especially since the beginning of the Industrial Revolution.

———. "Global Warming Differences Resolved." LiveScience (May 2, 2006). Available online. URL: http://www.livescience.com/environment/ap_060502_global_warming.html. Accessed September 23, 2008. Discusses how former discrepancies in temperature calculations between satellite and radiosonde data used for global warming analysis have finally been identified and corrected.

Silverman, Fran. "What's a Ski Area to Do as It Warms? Adapt." *New York Times* (January 13, 2008). Available online. URL: http://www.nytimes.com/2008/01/13/nyregion/nyregionspecial2/13skict.html. Accessed September 23, 2008. Looks at global warming's impact on the recreation industry.

Simons, Marlise. "Fungus Once Again Threatens French Cave Paintings." *New York Times* (December 9, 2007). Available online. URL: http://www.nytimes.com/2007/12/09/world/europe/09cave.html. Accessed September 23, 2008. Looks at the cultural effects of global warming and why the world's art may be in danger.

Stevens, William K. "On the Climate Change Beat, Doubt Gives Way to Certainty." *New York Times* (February 6, 2007). Available online. URL: http://www.nytimes.com/2007/02/06/science/earth/06clim.html. Accessed September 23, 2008. As research continues, there is enough evidence that scientists can no longer say the climate is not warming up.

Struck, Doug. "NOAA Scientists Say Arctic Ice Is Melting Faster Than Expected." *Washington Post* (September 7, 2007). Available online. URL: http://www.washingtonpost.com/wp-dyn/content/article/2007/09/06/AR2007090602499.html. Accessed September 23, 2008. Evidence that predictions of future ice loss may be too conservative.

Sturm, Matthew, Donald K. Perovich, and Mark C. Serreze. "Meltdown in the North." *Scientific American* (October 2003): 60–67. Current melting in the Arctic regions, what it is doing to the ecosystems, and how that will affect the rest of the world.

Than, Ker. "Animals and Plants Adapting to Climate Change." LiveScience (June 21, 2005). Available online. URL: http://www.livescience.com/environment/050621_warming_list.html. Accessed September 23, 2008. How some species have been able to adapt to changing systems.

———. "How Global Warming Is Changing the Wild Kingdom." LiveScience (June 21, 2005). Available online. URL: http://www.livescience.com/environment/050621_warming_changes.html. Accessed September 23, 2008. Looks at ecological issues among mammals, fish, insects, and other types of wildlife.

———. "The 100-Year Forecast: Stronger Storms Ahead." LiveScience (October 13, 2005). Available online. URL: http://www.livescience.com/environment/051013_stronger_storms.html. Accessed September 23, 2008. Presents the concept of an 'enhanced' water cycle and more severe storms as a result of global warming in the future.

———. "Polar Meltdown Near: Seas Could Rise 3 Feet Per Century." LiveScience (March 23, 2006). Available online. URL: http://www.livescience.com/environment/060323_ice_melt.html. Accessed September 23, 2008. This article explores the implication of sea-level rise as a result of the melting of polar ice caps.

———. "Global Warming Weakens Pacific Trade Winds." LiveScience (May 3, 2006). Available online. URL: http://www.msnbc.msn.com/id/12612965/. Accessed September 23, 2008. Outlines why the trade winds are weakening as a result of global warming.

Tierney, John. "In 2008, a 100 Percent Chance of Alarm." *New York Times* (January 1, 2008). Available online. URL: http://www.nytimes.com/2008/01/01/science/01tier.html. Accessed September 23, 2008. An overview of why action needs to be taken immediately to deal with global warming on a global scale.

Vergano, Dan. "Global Warming Stoked '05 Hurricanes, Study Says." *USA Today* (June 25, 2006). Available online. URL: http://www.usatoday.com/tech/science/2006-06-22-hurricane-blame_x.htm. Accessed September 23, 2008. Presents the reasons why climate scientists believe that global warming helped fuel 2005's destructive hurricane season.

Wald, Matthew L. "Cleaner Coal Is Attracting Some Doubts." *New York Times* (February 21, 2007). Available online. URL: http://www.nytimes.com/2007/02/21/business/21coal.html. Accessed September 23, 2008. Explores the use of new technology called 'gasification' to operate new coal plants and whether it will be as environmentally friendly as originally thought.

———. "Study Details How U.S. Could Cut 28% of Greenhouse Gases." *New York Times* (November 30, 2007). Available online. URL: http://royaldutchshellplc.com/2007/12/01/the-new-york-times-study-details-how-us-could-cut-2 8-of-greenhouse-gases/. Accessed September 23, 2008. Outlines how the United States could reduce the greenhouse gases it generates at a reasonable cost with only small technological innovations.

Walsh, Bryan. "The Fire This Time." *Time* (November 5, 2007): 32–43. Wildfires in California and their connections to climate change.

Warburton, Louise. "No Where Else to Go—Climate Changes and Parrots: Can they Adapt to Survive?" *Bird Talk* (January 2007): 22–85. Discusses the ecological impact global warming is having on the world's bird population.

Williams, Gisela. "Resorts Prepare for a Future Without Skis." *New York Times* (December 2, 2007). Available online. URL: http://travel.nytimes.com/2007/12/02/travel/02skiglobal.html. Accessed September 23, 2008. Discusses the economic ramifications of global warming on the tourist industry.

Williams, Jack. "Drilling Uncovers Past, Maybe the Future." *USA Today* (January 23, 1999). Available online. URL: http://www.usatoday.com/weather/resources/coldscience/adril.htm. Accessed September 23, 2008. This article reviews what climatologists have learned about climate change through the study of climate from the past via ice cores.

———. "Greenland's Ice Tells of Past Climates, Maybe Ancient Life." *USA Today* (August 30, 2004). Available online. URL: http://www.usatoday.com/weather/resources/coldscience/2004-08-30-n-grip-main_x.htm. Accessed September 23, 2008. This article reviews the importance of paleoclimatology.

Woollard, Rob. "California Wildfires Force Mass Evacuations." *Discovery News* (October 23, 2007). Available online. URL: http://dsc.discovery.com/news/2007/10/23/california-wildfire-print.html. Accessed September 23, 2008. Links the recent tragic wildfires of California and the American Southwest with drought and global warming.

Zimmer, Carl. "Migration Interrupted: Nature's Rhythms at Risk." *New York Times* (January 1, 2008). Available online. URL: http://www.nytimes.com/2008/01/01/science/01migr.html. Accessed September 23, 2008. Discusses migration corridors and how global warming is affecting them.

WEB SITES

Global Warming

Climate Ark. Available online. URL: www.climateark.org. Accessed October 23, 2007. Promotes public policy to address global climate change through reduction of carbon and other emissions, energy conservation, alternative energy sources, and ending deforestation.

Climate Solutions. Available online. URL: www.climatesolutions.org. Accessed October 23, 2007. Practical solutions to global warming.

Environmental Defense Fund. Available online. URL: www.environmentaldefense.org. Accessed October 26, 2007. An organization started by a handful of environmental scientists in 1967 that provides quality information and helpful resources on understanding global warming and other crucial environmental issues.

Environmental Protection Agency. Available online. URL: www.epa.gov. Accessed October 26, 2007. Provides information about EPA's efforts and programs to protect the environment; it offers a wide array of information on global warming.

European Environment Agency. Sponsored by the European Environment Agency in Copenhagen, Denmark. Available online. URL: www. eea.europa.eu/themes/climate. Accessed October 26, 2007. Posts their reports on topics such as air quality, ozone depletion, and climate change.

Global Warming: Focus on the Future. Available online. URL: www. enviroweb.org. Accessed October 26, 2007. Offers statistics and photography of global warming topics.

HotEarth.Net. Available online. URL: www.net.org/warming. Accessed October 26, 2007. Features informational articles on the causes of global warming, its harmful effects, and solutions that could stop it.

Intergovernmental Panel on Climate Change (IPCC). Available online. URL: http://www.ipcc.ch/. Accessed October 26, 2007. Offers current information on the science of global warming and recommendations on practical solutions and policy management.

NASA's Goddard Institute for Space Studies. Available online. URL: www.giss.nasa.gov. Accessed October 26, 2007. This Web site provides a large database of information, research, and other resources.

NOAA's National Climatic Data Center. Available online. URL: www. ncdc.noaa.gov. Accessed October 26, 2007. Offers a multitude of resources and information on climate, climate change, global warming.

Ozone Action. Available online. URL: www.semcog.org/services/ ozoneaction/kids.htm. Accessed October 26, 2007. Provides information on air quality by focusing on ozone, the atmosphere, environmental issues, and related health issues.

Scientific American. Available online. URL: www.sciam.com. Accessed October 23, 2007. An online magazine that often presents articles concerning climate change and global warming.

Tyndall Centre at University of East Anglia. Available online. URL: http://www.tyndall.ac.uk. Accessed October 26, 2007. Offers information on climate change and is considered one of the leaders in UK research on global warming.

Union of Concerned Scientists. Available online. URL: www.ucsusa.org. Accessed October 26, 2007. This Web site offers a quality resource sections on global warming and ozone depletion.

United Nations Framework Convention on Climate Change (UNFCCC). Sponsored by the United Nations Framework Convention on Climate Change. Available online. URL: http://unfccc.int/2860.php. Accessed October 26, 2007. A spectrum on climate change information and policy.

U.S. Global Change Research Program. Available online. URL: www.usgcrp.gov. Accessed October 26, 2007. Provides information on the current research activities of national and international science programs that focus on global monitoring of climate and ecosystem issues.

World Wildlife Foundation Climate Change Campaign. Available online. URL: www.worldwildlife.org/climate/. Accessed October 26, 2007. Contains information on what various countries are doing and not doing to deal with global warming.

Greenhouse Gas Emissions

Energy Information Administration. Available online. URL: www.eia.doe.gov/environment.html. Accessed October 26, 2007. Lists official environmental energy-related emissions data and environmental analyses from the U.S. government. Contains carbon dioxide, methane, and nitrous oxide emissions data and other greenhouse reports.

World Resources Institute—Climate, Energy & Transport. Available online. URL: www.wri.org/climate/publications.cfm. Accessed October 26, 2007. This Web site offers a collection of reports on global technology deployment to stabilize emissions, agriculture, and greenhouse gas mitigation, climate science discoveries, and renewable energy.

INDEX